Cover illustration attributed to Howard Brodie.
See Figure 2 reference, page 87

Also by Francis Resta:
The WAR, My Memoir of World War II, 2003.

THE COMBAT VETERAN

VETERAN

and PTSD

& Help For The Family

Francis E. Resta

Second Printing August 2009

ISBN 978-0-557-05533-3

Dedication

This book is dedicated to combat veterans, men and women, returning from the Middle East wars, and to their spouses.

There is a special, heartfelt dedication to my granddaughter and her husband, a Marine who survived three tours in Iraq.

ACKNOWLEDGEMENTS

This book on the Combat Veteran and PTSD would not have been possible if the author had not had a number of experiences which must be acknowledged.

Dr. Melinda Keenan, Veterans Administration Hospital at Mather, California, who, through PTSD therapy and endless encouragement, gave much information and inspiration to the author.

The Sacramento Veterans Center team of counselors at 1111 Howe Avenue, Sacramento, who either provided the author with, or induced him to find, much of the information in this book.

The book, *WWII, A Chronicle of Soldiering* by James Jones, author of *From Here to Eternity,* is an excellent book about combat veterans. It provided the author with much more understanding about why a combat veteran is the way he or she is. It should be noted that Jones also wanted to show all the combat veterans' art created in WWII, lest it be forgotten. He collected 186 of those art reproductions from all countries involved in WWII, to show in his book. Although it is the best book on the realities of WWII that I have ever found, it's out of print, but must be available in some libraries. It was published in 1975 by Ballentine Books, Library of Congress # 74-27944, ISBN # 0-345-25428-7.

The author is forever indebted to Seanchan Owen for hours of thought-provoking philosophical discussions and of computer expertise to help prepare a more professional final version as well as the support and encouragement to get it done. I shall never forget.

My forever love for my lovely partner, life-friend, and personal fan club, Carol Bergen, will never be enough to properly thank her for the enduring and uplifting support as well as careful review to make this book a class act.

Also, I am grateful to live in a country where I can receive therapy and relief from PTSD by journaling my war experiences with the help of the Sacramento Veterans Center and Veterans Administration Therapists, and from that journaling to create this book, and, finally, to thereby receive the great reward of helping other combat veterans heal themselves.

Figure 1 -The End of The Trail, a bronze sculpture by James Earle Fraser

PREFACE

The well-known work called "The End of the Trail" by sculptor James Earle Fraser, shown on the facing page, depicts an exhausted Indian and his tired horse, beaten not only by weather but by life, perhaps finally getting home after a terribly wrong loss, a fruitless hunt, a family burial far away, or loss of comrades in a pointless war battle.

I keep the sculpture on my mantle to remind me of my experiences in WWII, which I never want to forget lest I stop demonstrating against war.

This book began as a talk I was asked to give at the Davis, California, Senior Center on Veterans Day 2008. Col. Ed Trautt, retired, WWII AAF, a Veterans of Foreign Wars comrade, asked me to talk about Veterans' benefits and I did, but I also wanted to tell my audience about what a returning combat veteran is like. The talk was so well received and the demand for copies so gratifying that the 30-minute talk grew into this book.

I hope that such information as developed in this book might help veterans coming home from Iraq or Afghanistan. This book is intended, as was the talk on Veterans Day, for returning combat veterans and for the loved ones who must interface with them and help them to become healed and productive citizens again. It is applicable whether you are a veteran returning from a recent theater of action or a long-time veteran wanting healing, never too late. It is particularly applicable if the veteran reads it with a VA therapist as therapy assignments.

Although this book was written from WWII perspective and experience and thus reflects those experiences which may be particularly meaningful to WWII veterans, it nonetheless is applicable to combat veterans coming home from any war. It is intended as a no-nonsense guide telling how the combat veteran gets PTSD (Post Traumatic Stress Disorder), what PTSD is, and what it does to the veteran.

Recently returned veterans may find it interesting to compare the wars they endured with WWII, which is portrayed the way it was, no punches pulled.

I've put the first part of my original talk on Veterans Benefits as Appendix A, *Veterans Benefits*, and Appendix B, *How Does One Go About Getting Veterans Benefits*, because they are applicable today for veterans coming home from the Middle East wars, and I believe they must be broadcast every way possible. Older veterans may find that many benefits are still available for any veteran.

CONTENTS

CONTENTS – Continued

AUTHOR'S NOTE

To eliminate awkwardness of style, the masculine gender has been used for most pronouns in the early chapters of this book, where most of the discussions reflect my experiences in WWII; most of the veterans I served with were men.

PTSD itself, however, knows no such distinction. No one should ever forget that women are most certainly included in discussion of PTSD, particularly in view of their significant service in the combat theaters of the more recent Middle East Wars.

Of course, the Nurses Corps, mostly women in all wars, saw the insanity of war through trying to patch up the wounded combat soldiers and watched many of them struggle with amputations or die. Also, when our armies were attacked, the nurses had to experience the same horrors under fire as did the combat troops. The more recent wars have both men and women medical corps persons undergoing those same experiences.

**

I saw that sacrifice when I was at a staging hospital, a series of tents in Liege, Belgium, after I was wounded in Germany. A German V1 "Buzz Bomb," probably intended for England, fell short and struck the hospital tent area, wounding nurses and some already wounded soldiers and killing some. Those wounded who could walk tried to give help and comfort to the screaming nurses.

INTRODUCTION

Universally, people most fear those things they don't understand, and Post Traumatic Stress Disorder (PTSD) behavior is one of the most tragic. This book will cast light on how the combat veteran gets PTSD, what PTSD is, and what PTSD does to the veteran. Learning about PTSD is the only way a combat veteran can cope with this disorder and the only way a loved one can find a pathway to tolerance and loving communication.

This book is intended to be useful to five main areas of interest:

1. It is a clinical handbook for the psychology of PTSD. It provides important information for both the therapist and the veteran. The therapist can also use it for a workshop platform, assigning sections or paragraphs for the veteran to read in session and discuss feelings and experiences while leading the veteran to denial-free understanding.

2. It is a combat veteran's guide to understanding his or her PTSD. The veteran can get new meaning each time he or she reads through the book. Reading several times those parts which seem especially gripping to the veteran will provide much healing. By applying the lessons learned from this book, the veteran can become conscious of his or her behavior, not as that of a bad person, but as a damaged person who can become whole again.

By reading this book and especially if the veteran journals his or her feelings about each section or paragraph, an awareness of PTSD and its affect on the veteran can grow into real therapeutic gain in reducing the affect of PTSD. Further, another significant benefit for the veteran in the journaling, particularly for Chapter IV, *What is the Combat Veteran Like?*, with its accounting of personality characteristics caused by or exacerbated by the veteran's combat experiences, the veteran would be well equipped to make self improvement changes.

The veteran would recognize which of his or her personality characteristics are reflected in this book, a healing process in itself. Realizing changes in his or her personality which the veteran was not aware of, the veteran could then focus on consciously choosing to change those which he or she might consider undesirable.

3. It is a guide for family or friends of a combat veteran to find greater understanding of the veteran's problems and behavior. Reading it will provide more sensitivity and compassion to keep trying to find connection. The veterans' loved ones can find new perceptions and meanings with which deeper bonding can be constructed.

The veteran and the loved one could read this book together and talk about their feelings as they read along. That must be done with caution: anguish on both sides might be provoked, and if that happens, you may need counseling together. However, both could read it together with the premise that it should not be a revealing of all the truth about what the veteran did in the war –that would be for a therapist to lead. And the reading should not be to overwhelm the spouse with the agony and horrors discussed in this book or to hear the veteran's similar experiences. But the healing for both parties could be significant through understanding the healing that both might need. The resulting bonding would carry them through many of the trials that marriage will bring.

4. For WWII veterans, this book provides stories with which he or she can relate and find bonding through them. For veterans of other wars, the WWII background in this book will contrast the different war and provide insight into what factors are the same and what are different. It will help any veteran to appreciate his or her experiences as well those different experiences of the WWII veteran.

Further, for the WWII buff, it provides true, no-nonsense depiction of the way WWII really was. If you are looking for the glory of war, don't bother to read further.

5. If you are not a combat veteran but your childhood was continual trauma and abuse, you also might have PTSD and will find yourself in this book, particularly Chapters III and IV. By reading those chapters you can experience the same wonderful healing that a combat veteran can get.

**

In this book, the primary focus is the depiction of the causes of PTSD and its affect on the combat veteran carrying the burden of grief with a huge sense of loss, of guilt, and of shame. Throughout the book I will be talking about those burdens sometimes with text description and sometimes with anecdotes from my experiences as a combat veteran. My intention is to help the non-veteran reader to **feel** what it's like to be a combat veteran, to have been there and experienced it all. Although many books have been written with that intent, few succeed.

Chapter 1
EVOLUTION OF A COMBAT SOLDIER

World War II (WWII) – An Outlook

To open this discussion about combat veterans coming home, I want to draw your attention again to the sculpture "End of the Trail," facing the PREFACE, and the utter fatigue that it depicts.

To me it reflects how we combat infantry in WWII felt most of the time. There was no day or night in combat; no scheduled time to sleep or to go to bed; no bed to go to. We usually just slept lying on the ground or in a slit trench[1] or propped up in a foxhole. We were often exhausted enough that we learned how to fall asleep for a few seconds even while standing, for instance when a moving column stopped and started as we were moving along not under fire. Some could even catch a few seconds of sleep while walking; one soldier would see the guy in front sort of fall to one side in slow motion for a moment before waking up. The soldier behind him might reach for the stumbling guy's shovel on the back of his pack, trying to keep him from falling.

In combat, there was no set time to eat, sometimes no food to eat. Combat didn't stop for holidays or weekends or even at night when there were patrols and mortar firing to do, and the artillery never stopped. Night was a time we could move, undetected, so there was work to do every night. A part of every night was the complete darkness; to

---

[1] A foxhole is just deep enough and wide enough to sit in and to not be exposed to enemy fire. It can be made deeper or wider, if it will be "home" for a while and even include a 'shelf' to sit on. A slit trench is as long as your height and just deep enough or wide enough to be able to lie down in it without being exposed to enemy fire or shrapnel. It was used at night if, during an attack, you needed to "hide" temporarily to sleep without shrapnel or small arms fire accidentally hitting you.

show a light, even lighting a cigarette cupped in your hands would bring immediate German fire, perhaps even artillery. When there was no moon, utter darkness engulfed you and you just waited for dawn. Of course, there was no TV or even radio. Just endless time. Guard duty was the loneliest, not having any one to talk to, not that there was much talking to do unless you were in a dugout where your voice wouldn't carry (in wartime's quiet night time, even a whisper can carry a half mile!)

No Sunday to rest before Monday back-to-work. We slept a few hours when we could and where we could, always interrupted before we could get into a deep sleep. The pieces of sleep seldom added up to six hours in 24, which hardly gave us rest from the intense physical activity we had to do. Everything we needed had to be carried on our backs: weapons, ammo, food, overcoat, gas mask – adding up to 70 or 80 pounds. You got used to the weight –it was always there.

We seldom ate a meal, but snacked on rations when we could, seldom adding up to one full meal in a day. When it rained and if we didn't happen to be in a sheltered place, we just got wet. Sometimes for days, wet through and through, all our clothing, our shoes, everything. Living and walking in mud was just like any other day. In the winter, our feet got frostbite[2]. Sometimes in the winter if we were not moving, we dug holes over and over, just to keep warm. Our heavy wool overcoats were the only winter covering we had, and we were often chilled enough to actually feeling pain. The war went on day after day, week after week, month after month. Endless, with no end-of-the-trail in sight.

My unit, Company B, 407th Infantry, 102nd Infantry Division, was on line in Germany one time for 218 days, pressing slowly but relentlessly into Germany from

---

[2]   Two-fifths of the casualties at one hospital, (Southampton, England, where I was sent when wounded), were frostbite; many had lost toes; some had lost feet. I was probably sent there because I had frostbitten feet as well as shrapnel wounds.

Holland, up to and the assault through the Siegfreid Line, crossing the Roer River, and finally crossing the Rhine River at Wessel. That 218 days was forever. Then we made the dash across Northern Germany to the Elbe River, where we waited near Stendhal for the Russian Army to get to the Elbe as the German Army collapsed.

It was an endless time for many, although those who got wounded as I did, were grateful to get a break from war while our wounds were healing. In the hospitals we slept on sheets in an actual bed, flirted with nurses, ate three meals a day, and could sleep, blessed sleep, as many hours a day as we wanted, all for the first time in months. When our wounds were healed, we were sent back into combat.

Some veterans had multiple Purple Hearts (each awarded for one combat wounding). If we didn't lose a limb or some other necessary part and would therefore be sent home, we always were sent back into combat after healing. The "million-dollar wound" was losing fingers or toes, which wouldn't incapacitate one after the war for most of life's needs, but would result in being sent home because of loss of capability to be a combat soldier.

The movie *The Big Red One* with Lee Marvin, about the 1st Infantry Division, is, in my opinion, one of the best to depict the endlessness of combat. Those men who survived from the beginning of the American WWII ground offensive in Africa, then through Sicily, Italy, D-Day in France and through Germany, are the true heroes of WWII. In all those three years, few of the men in the Big Red One had opportunity for a break. Many didn't survive. Many were wounded multiple times and were sent back into combat over and over. Many were crippled by the effects of combat on their minds and bodies. For them, the war just went on and on forever. When they got home, handicapped with severe PTSD, the war just went on and on in their heads and emotions, as we will see in later chapters.

The Soldier Learns Combat Anger

I can tell you that a combat veteran coming home is angry, almost overwhelmingly, and quick to lose temper with himself, with others, and with circumstances. He is aware in some part of himself that he survived the war and therefore, somehow, unconsciously, believes himself to be invincible and due for the good things in life. He has no patience with roadblocks in his life and is prone to seemingly unprovoked emotional outbursts, uncalled for aggression, or sudden and seemingly inexplicable, deep depression. But his anger after discharge is often hidden, like a dog that can attack without a warning snarl. The veteran often hides his anger as he does with all his emotions, as we will see in this book.

Often the veteran thinks his anger is generated by his being betrayed by his country sending him into harms way in such an insane environment, doing things he cannot forgive himself for and feeling he betrayed his buddies by surviving when they didn't. But the veteran doesn't realize that the actions and experiences of combat weren't the causes of the anger, but instead it was the other way around. The anger he had been trained to use for focus was what enabled him to live through combat and do the things he did.

Anger and loud aggression are a large part of the training for combat and mask a boy's sensitivity to those events and circumstances that he must quickly and painfully give up entirely to become a military soldier. His individuality, his belief in fairness and justice, his faith in the goodness of people, and any expression of naive sensitivity– all must go. The reader will see, as the detail of this book unfolds, how anger ruled the combat veteran's life. The young soldier learns, almost by accident, that the anger and aggression that wells up in him as his boyhood persona becomes firmly and relentlessly squelched, is acceptable when focused against the "enemy." He is constantly rewarded for expressing anger and aggression. It is a daily part of his training as a recruit; he is constantly

4

yelled at not only critically and demeaningly but also prodded to make noise and yell and to show anger while, say, stabbing an enemy dummy on a training course.

Learning to Kill

An important part of the recruit's training, both in terms of learning how to survive and also in terms of its hidden, terrible affect on the combat veteran, is the instruction and drills on how to kill –with bullets, grenades or other weapons, knife, bayonet, rifle butt, by hand (Judo, in WWII), including much shouting and running, thrusting bayonets at realistic dummies (some containing red dye to squirt out when jabbed, and slitting throats. The recruit was yelled at over and over during the killing drills to hit or bayonet harder and with all his strength. The drills were endless, day in and day out, and become accepted by the soldier as just a reflex, to kill someone. Sadly, the soldier begins to feel invincible, with the feeling that he could easily kill anyone –"no one is going to get **me**!" A veteran fiend once asked me, "you mean the training cadre were messing with our minds?" Unfortunately, the answer is "Yes!" The only way countries throughout history have been able to win wars is to create combat veterans. That means more than just training and equipping men to fight. It was the moral act of creating a person that could perform in the insanity of war, whatever that would take.

Realize that the training both as a recruit and by baptism of being under fire didn't just train us **how** to kill; it taught us to kill. As Ernie Pyle narrated in Ken Burns' film *The War*, (Episode Two) of the American combat soldier in early fighting in Africa:

"The most vivid change in our men (as they were finally pushing the Germans and Italians out of Africa), is the casual and workshop manner in which they now talk about killing. They made the psychological transition from the normal belief that taking human life is sinful, over to the new professional outlook where killing is a craft to them now.

5

There's nothing morally wrong about killing; in
fact it is an admirable thing." (Ernie Pyle)
**

I don't know any combat veteran who doesn't
remember his first killing. You could dramatically say that
that vision is burned into his soul, seeing the enemy soldier
just suddenly sort of slump into himself and fall or roll
over. But most of us just wish we could forget.

The WWII documentary directed by Ken Burns is
mostly actual war films of WWII and is one the most
realistic of war films I've seen. It shows war the way it
really was: the mistakes, the sometimes-sheer luck, the
heroism, the suffering, the endlessness, the accomplishment
and victories, and how war plays no favorites; everyone has
an equal chance to die or to survive. (You can tell real war
action footage from Hollywood stunt-acting by the recoil of
hand weapons –there is no recoil in Hollywood film.)

A part of Infantry training also included endless hours
of "indoctrination" in how inferior the enemy is, how the
"Japs" are not very smart and can't invent anything but only
copy from another country, or that they can't see very
good, and how the "Heinie" only knows how to follow
orders –he'll just stare at you as you kill him. The training
included film after film of atrocities performed on the
Chinese and of islands overrun by the Japanese and the
European countries overrun by the Germans, until you
actually hated them viciously and were determined to seek
them out and kill them.

A New World – Changing Culture

The reader probably isn't surprised that the training of a
person for combat would have to be severe to succeed and
result in survival. However, the intensity of combat is not
only a primary affect that causes PTSD, but also is a factor
that changes a veteran in ways one wouldn't expect. The
cultural sociologist would see it as a simple process: when
a person changes cultures he or she becomes a different
person, over time. We all have had many changes in

culture throughout our lives. Starting perhaps in the first few months of life when Mommy or Daddy went back to work and your life changed markedly. Or when you started kindergarten or high school, or your first job or when you first went away from home, off on your own. Or going to college, getting married, first home, first child, moving to a different city, etc. It goes on and on, the changes to our lives.

Sometimes it takes months to get used to the new culture, and many books have been written about cultural shock for the various changes in our lives and how to cope with them – a sort of boot camp for the important changes. But realize that these changes in our lives are all **within** our culture. To go to another country, or to a developing country, or even an aboriginal culture, or living with gorillas or bears or wolves. Those are reaculturations wherein a person is drawn into vastly different lifestyles that would test the best of us!

The movie "Dancing With Wolves[3]" illustrates the reaculturation of Lt Dunbar (Kevin Costner), who joined the culture of the Lakota Indian Tribe, after he had already changed culture to be a lonely, sole military member at the Indian lands' military post, and before that also having changed culture from growing up a civilian to be a cavalry officer in the Civil War. It may be possible to become so many different people, but I wonder at his sanity. The movie did effectively depict his having become so completely a Lakota that he reflexively killed (with his Indian friends' help) all of the U.S. Army patrol that had captured him as a military traitor.

The WWII recruit who had been changed from a moral, naïve believer in goodness, fairness and justice into a fighting man, taught to fend for himself in a terribly hard way of life and to gladly kill another human being, was a

---

[3]   According to internet's Wikipedia, *Dancing With Wolves* was selected in 2007 for preservation in the United States National Film Registry by the Library of Congress as being "culturally, historically, and esthetically significant."

changed man in ways that the rest of this book endeavors to explain, along with what that changed veteran is like when he or she comes home. Living in the entirely different world of combat which is simply unknown to those who haven't lived it, has changed the soldier into an entirely different person that the people back home simply can't and don't know anymore, as we will see in Chapter IV of this book.

**

Another aspect of training was a growing sense of pride from being a part of a mighty military machine. Even in combat, a part of you somehow felt safe even though another part of you was terrified into numbness, at least at first. In Ken Burns' film *The War,* one of the veteran narrators said it so succinctly:

> "...you knew you were in great danger but you somehow felt safe and that you were a part of this powerful group....you would feel the power of the military."

I recall that in the 102nd Division farewell parade at Ft. Dix, NJ, we in Co. B of the 407th Infantry Regiment, were lucky to see the parade from the vantage point of being at roughly the center of the long marching line, behind the 405th and 406th Regiments, and at the front of the 407th Regiment and then all the ancillary units spread out about a Regiment's size at the end. After marching across the back of the parade field then turning to pass in review to march the other direction across the front of the parade field, we could see the entire 102nd Division, 18,000 men and many pieces of artillery, trucks, light tanks, ambulances, etc, moving in both directions as we looked between the two long moving lines. The sight was hugely overwhelming, and with the U.S. flag shortly in front of us, we just **knew** we were destined to make history and many of us had tears of emotion in our eyes.

I had the same feelings at the start of the attack on Welz, Germany. When we rose up out of foxholes in a several mile long line in the beet fields, I looked to the right

8

and saw the thousands of men of the 407th Regiment, starting the WWI-type charge across the fields. The feeling of being a part of a military might, of being invincible, was overpowering. Some of the men started yelling, like in Civil War movies' charges. *(Keep that thought in mind when you read at the end of this chapter what happened at Welz after that magnificent sight.)*

The Fight or Flight Instinct

The focus in the recruit's training is to mold the "fight or flight" reaction natural in all of us so that only fight, never flight, is the sole response the soldier will make in combat. That human instinct to flee if needed is fiercely guarded by our whole psyche, and the tough job of the training cadre is to change the human reaction into the combat reaction to never flee, to always fight, to always charge recklessly and to never give up. It is another of the subtle but powerful changes in making the recruit a wholly different person in a wholly different world.

I remember when, even as late in training as waiting at Fort Dix, New Jersey, for embarking to sail across the Atlantic for combat, we in Company B learned a somber lesson about fighting. Two of our guys were turned over to our Company B Commanding Officer (CO) by the MPs policing the nearby town. The two big B Company men were charged with drunken and disorderly conduct and fighting with the MPs. The CO was directed by the MP officer to give them "company punishment," a form of military justice in which the CO had complete freedom to mete out any punishment which he might decide was appropriate, just or not. The Officers answering to the CO and the NCOs answering in turn to them, had the same privileges over us; we knew that system from living under it for our years of military service. An officer can decide to mete out company punishment in any form he wishes any time he wishes –no trial, no defense, no questions asked or answered.

The CO had the company fall out to watch the punishment, and then ordered the two returned men to each dig a 6x6x6 –a 6ft. wide by 6ft. long by 6ft. deep hole, quick-time. Although one can start digging quick-time, it soon becomes an agonizing, brutal action even for fit men[4].

As we watched the two suffering men, the CO explained that this punishment wasn't as the MP officer had asked, for disorderly conduct and fighting with the MPs. Instead, the CO wanted us to realize that it was for not beating up the MPs, killing them if they had to, but to never surrender to the enemy. I'll never forget the CO shouting: "My God, you're the fucking Infantry! But you're not ready yet for combat and you won't be until you never surrender alive and your only thought is to kill the fucking enemy!"

**

Hiding the disappointment, heartache, and disillusion-ment that the boy must no longer show becomes an automatic response, a defense against feelings. It is the start of the training to suppress feelings, a training to be able to "numb out" in combat, as we will learn about in Chapter III. It eventually seems to make sense to the man as he becomes a combat soldier. By then, he has successfully masked memory of the boy and doesn't remember the training that taught him to do so.

After getting into combat, the horrors and what he does as part of war prompt him to seek a reason for his anger and feelings of betrayal. It is easy for him to focus on those who sent him into combat, the people at home; they are now, dimly, somehow, the enemy. Not the family, of course, but "them" --the politicians, the draft board, the corporations. You know, "THEM."

---

[4]   Digging 6x6x6s was training for survival: when under fire we had to quickly dig a fox hole or even a slit trench, and believe me, when you are motivated by being under enemy fire, you find the energy to dig quick-time or even double quick-time!

Sometimes during combat, the Regimental or Division brass became the "enemy." The rear echelon officers were seen as not always making good tactical combat decisions and were then directly the cause of tragic losses, as illustrated at the end of this Chapter.

Becoming a Combat Soldier

The book *WWII, A Chronicle of Soldiering* by James Jones, author of *From Here to Eternity*, is an excellent book about combat. James Jones was in a combat unit and wounded early in the South Pacific War. It's a book about how war really is, the fatigue, the horror, and the endlessness. The following excerpt describes the "evolution" of a combat soldier:

James Jones:

What is it, really, this EVOLUTION OF A (COMBAT) SOLDIER? ….I've been talking about it all through this book, but I'm not sure I can explain or define it. I think that when all the nationalistic or ideological propaganda and patriotic slogans are put aside, all the straining to convince a soldier that he is dying *for* something --it is the individual soldier's final full acceptance of the fact that his name is already written down in the rolls of the already dead.

Every (combat) soldier, if he follows far enough along the path that began with his induction, must, I think, be led inexorably to that awareness. He must make a compact with himself or with Fate that he is lost. Only then can he function as he ought to function, under fire. He knows and accepts beforehand that he is dead, although he may be walking around for a while. That soldier you have walking around there with this awareness in him is the final end product of the EVOLUTION OF A (COMBAT) SOLDIER.   (Jones 43)

We in the Infantry in World War II were told over and over in training and in early overseas briefings, that if we did our job none of us would live to go home. When finally starting in combat, it didn't take long to realize that that was probably true. When those of us who survived **did** get home, we felt we had somehow betrayed those buddies who "did their job" and didn't make it. We couldn't shake the deduction that somehow we failed to do our jobs, or perhaps we are cowards and that's how we survived, and we felt angry and betrayed by **some**thing.

But being betrayed was something the boy would have felt. Perhaps that part of the boy wasn't wholly suppressed in all of us veterans. Some more than others have that boyish sense of betrayal and do seem to need a good shaking and to be asked, "When are you going to grow up? You have to accept what is happening to you." I often felt betrayed, and even getting wounded was a sort of betrayal –it wasn't supposed to happen to **me,** just to the others. The wholly hardened combat veteran seldom felt betrayal; he seldom felt anything.

In Ken Burns' magnificent documentary film *The War*, in Episode One, a veteran is trying to explain what that hardening from the terror of being under heavy fire felt like:

" Some men could take it and some just could not take it. The sheer terror of knowing that the next one is going to have your name on it. And that just goes on and on and on. You get a strange feeling in which you seem to become detached. And you just think 'well, maybe this will end and maybe it won't, and maybe we'll all be blown up and maybe we won't, but who cares'. You had to sort of live with it. ...you will either survive or you will not, so there's nothing you can do and you just take it" (Episode One)

And in Episode Five of Ken Burns' *The War*, a veteran is recounting how a soldier, early in combat, is beginning to understand his insignificance in the uncompromising horror of war:

"Slowly the reality of it all formed in my mind.
We were expendable –it was difficult to accept.
We come from a nation and culture that valued
life and the individual. To find oneself in a situation
where your life seems of little value is the ultimate
loneliness." (Episode Five)

As noted frequently in Ken Burns' *The War,* being afraid in combat was usual and puzzling to the soldier, and got worse, although more suppressed, as combat went on and on. Most Veterans came home with deep feelings of guilt for having often been afraid and the feeling therefore of being a coward. No amount of TLC or praise would make that feeling go away. Being called a hero was, to most combat veterans, an insult; it could only be heard as sarcasm.

"Friendly Fire"

Even combat had tinges of betrayal at times, for instance when so-called "friendly fire" killed us just as completely as did German fire. The reasons for friendly fire in WWII are legion, and at the time they happened, media coverage was scant and probably suppressed, different from today. In WWII the military didn't have to make an "investigation" of the reasons for friendly fire as they are compelled to do in this century's wars.

There were many casualties from friendly fire in every assault and major offensive in WWII; it was just a "normal" part of war. Many books have documented friendly fire incidents and casualties in the Pacific landings, in the North African assaults, in the landings in Sicily and Italy, in D-Day confusion at Normandy that alone had several major confrontations where some U.S. Divisions were blindly fighting other U.S. Divisions, and many more instances.

Sometimes, when in combat and it seems that you've been abandoned and have to stay where you are and keep fighting (as at Bataan and Guadalcanal), you feel that the mounting casualties are sort of caused by friendly fire in the sense that the generals have forgotten you and are not doing anything to rescue you; you can't know that everything possible is being done to relieve you, but it's just tough luck. At Guadalcanal, the marines were stranded with no food and limited light weapons and ammunitions, and minimal supplies. They nearly starved to death several times, and had to live off the land for six weeks and all lost 20-25 pounds. The supply ships had been chased off by the Japanese Navy shortly after the landings, before the Marines could off-load even the meager 10 days' supplies they had brought with them in the ships, or the heavy weapons and ammo they needed for a sustained fight with the Japanese. As the Marines fought on against overwhelming odds and misfortune, there was no communication with the Navy, and supplies and reinforcements couldn't get through for six terribly long weeks[5]

**

The reasons for friendly fire in WWII were many: poor command strategy, poor soldier performance, poor soldier and command training, poor equipment, poor intelligence, etc., and often just bad luck. One might think it would be appropriate justice to make the generals account in a court of law for their mistakes and misguidance, which often caused unnecessary casualties. But that would mean that every Colonel, Lieutenant, Sgt, and soldier would also have to account for all **his or her** mistakes, too! But being a professional commander like Eisenhower or Bradley didn't mean they could possibly humanly control the absolute insanity and unpredictability of combat. As WWII dragged on, and even seasoned veteran commanders and soldiers were still making mistakes, it was not possible, except in Hollywood movies, for right to triumph and the wrong to perish, or for combat skills to be learned and always

---

[5]   Ref. Ken Burns' film *The War"*, Episode One

effective. The combat soldier of any rank began to feel that his belief in himself as a good person and of good judgment was inaccurate; he began to fault himself in a primal way that would stay with him throughout his life.

It shouldn't be a surprise that the traumas of combat were particularly painful when you learned that your horrible experiences and loss of buddies were caused not by the enemy but by our own forces or blundered tactics, or from feeling expendable such as at Guadalcanal[4]. The traumas and losses were heightened with that sense of betrayal.

**

For example, after pushing into Germany nearly up to the Roer River, the 407th Regiment of the 102nd Infantry division was literally slaughtered in the attack on Welz where I and many others were wounded and more were killed than in Pacific War beach landings. Earlier in this chapter I related how glorious it felt to be part of the invincible American war machine, as we rose from the foxholes in the beet fields before Welz, and started a full Regimental charge across the beet fields and up the hill to the town. But that was not how it ended

When our 102nd Division newsletter, *Ozark Notes*, in 1996 published an article about how great our military intelligence had been during WWII combat, I responded with a "Letter to the Editor" about the Welz fiasco, where everything that could have gone wrong, did –Murphy's Law was hale and hearty that day, November 30, 1944, -- and basically because of bad intelligence.

I also wrote the letter to protest another betrayal I felt, the attempt by some veterans, even combat veterans, to gloss over the realities of war and to try to find some sort of glory in it. I think that veterans trying to believe in glory is part of the attempt to forget our own inadequacies and self- deprecation. If we could believe that someone was heroic, then we weren't all bad. Some veterans in my 102nd Division, including myself, felt inadequate in having a relatively short and relatively easy exposure to combat, compared to some of the Pacific Marines and some of the three-year ETO Infantry veterans.

In my letter I detailed what really happened to us that day at Welz, Germany, an entirely different story than the Army news release to the newspapers or the implicit story in the *Ozark Notes* article.

**

My letter:

June 14, 1996

Letter to the Editor
OZARK NOTES, 102nd Div. Assoc.

Dear Sir or Ms:

This letter is in response to the article on the Cover Page of vol. 48, No.3 of the OZARK NOTES, April/June, 1996, titled *"Ozarks Standing Tall"*, a critique by Col. Douglass Johnson. For that evening's talk, the article noted that Col. Johnson did his homework in researching military history archives.

Let's not get so carried away with finding glory that we rewrite history. While it is comforting to hear that military historians are giving Ozarks kudos for excelling in "...our emphasis on intelligence," all the good 'B' Company, 407th men who were casualties and fatalities in the bloody attack to take Welz on the Roer River on November[6] 30, 1944, would have a pretty blank look hearing that (story). The 407th was shattered in that attack because intelligence

---

[6]   For some reason, this letter to the *Ozark Notes* as sent showed the date of September 30th, for the attack on Welz, not November 30th, the correct date. Suffice it to say that November 30th, because of its horror, has always played with my mind.

didn't know that a Panzer Division was waiting for us there. We sustained enough direct casualties to be stopped cold, and the reserve 406th went through us to continue the attack. I was one of the lucky wounded (not among the dead) in that attack, and the 407th was so shattered that I went through the 406th Aid Station --no one knew where the remnants of the 407th were.

During the attack, we not only were pinned down out in the beet fields with direct "grazing" fire from the Tiger 88s, but when we finally, after heavy casualties, got moving across the fields, we came under our own D. Co. heavy machine gun and 81mm mortar fire in the woods around the town of Welz, then under our own 90mm (and 105mm) Artillery fire when we got into the town, and ultimately were bombed and strafed by our own Air Force, which forced us out of Welz after taking it briefly. The wounded that day didn't get out until the 406th rescued us after dark, since the Germans were shelling ambulances and shooting Corpsmen.[7] It doesn't surprise me to have heard that a

---

[7]   The reason everything went wrong that day at Welz was two-fold. Firstly, the basic failure of intelligence to know that a German Panzer Division was waiting for us. That resulted in a WWI-type Regimental full frontal charge of just men out in the open against hidden tanks. If that had not been the case, the other things that went wrong wouldn't have been so devastating. To make it worse, all forms of our communications went out –walkie-talkies, back-pack radios, even hand signals in the smoke screen that engulfed us.
   Secondly, the panels weren't pulled in, probably because the Germans shot the GIs responsible for doing that. One of the unusual and heroic jobs in WWII not talked about much, was the putting out and taking in of large, plastic, brightly colored panels which designated, for Air Force pilots, where the front lines were --anything on the German side of the panels was fair game for killing. The job of pulling panels in the middle of an attack had about the same life expectancy as an outnumbered and ambushed combat patrol.
   Further, the German intelligence must have known we were going to attack across those beet fields, because the only way the tiger tanks and 88mm cannon could have been lined up so accurately to provide grazing fire so effectively was if that had been carefully planned.

17

War Correspondent with us at Welz said that although he'd been at Salerno and Anzio, he never saw such carnage. I understand that a Corporal White was posthumously awarded a Silver Star for shooting down one of the P-47s dive-bombing and machine gunning us that day, with a Jeep-mounted 50 cal. Browning (one of 'B' Co.'s innovations).

It should be noted that the *NEW YORK TIMES* article about that attack (which clipping my Dad sent to me) stated what the military historians wanted the public to know: The Times wrote that the last of the Siegfried Line had fallen after the 102nd Div. attack against only a few machine gun "nests" holding out, and only a few U.S. casualties resulted. It is *interesting* [my italics] that those words were almost verbatim what we were briefed before the attack: the "intelligence expectations" we were assured we would encounter.

I'd suggest that doing homework at military history sources isn't the way to get the facts. And, anyway, why are you needing to find glory? I "stand tall" because I am a caring and loving person, and sure as hell not because I participated in the insanity of WWII combat. Do you think the 407th Commander, who was relieved of command and sent home after Welz stands tall because of WWII experience? Hopefully he has healed and stands tall because of who he is.

Sincerely,
Francis E. Resta

The *Ozark Notes* never published my letter.

VE-Day

Figure 2 -VE Day

Chapter II
GOING HOME and READJUSTMENT

End of the War

The end of the war was a mixture of joyful cheering and new agonies. On VE-Day (Victory in Europe), May 8, 1945, we in the 102nd Division in Germany were atypically cheerful. Finally, the war was over and we were headed home! The long faces and cheerless, hollow, expression-less eyes were starting truly to smile, and we were full of kidding and doing practical pranks to each other. No more dying. No more killing. No more endlessness.

Much of our feeling of relief is reflected in the facing picture, which is a copy of a sketch I saved from a *Stars and Stripes* wartime periodical, showing the battle-weary combat soldier throwing down his rifle –thank God, the war's over! I had it framed and it hangs on my wall at home, again to remind me of what war really was.

But, as fate would have it, the 102nd Infantry Division was slated for redeployment next month or two to the Pacific for the assault on Japan, and so we began to be re-outfitted for that. But our emotions were not re-outfitted. We were reminded that we were in the Army for the duration and were soon to be shipped to the Pacific and go back into combat. So we took advantage of the respite by getting drunk and playing endless rounds of Pinochle, "21," and chess, while we waited and did some half-hearted training. We knew, though, that those of us who had learned how to survive against German tactics and weapons would have to learn all over again, by trial and error again, how to survive against the Japanese. But in that trial and error learning, how many more buddies would we lose? All of us, probably.

**

You need to realize that although the stateside training taught us how to kill and to want to kill the enemy, when the green soldier was first exposed to combat, the terror and stupefying shock was overcome only by seeing what the nearby experienced combat soldier was doing and trying to emulate him, or do what the experienced officers and NCOs yelled at him to do. But in North Africa and the early Pacific landings or when a whole new green division (like the 102nd) went into combat without anyone to show the way, the horror was almost always catastrophic. The intensity of combat simply stopped you cold if you weren't nudged to move. In North Africa, the terrified, green units simply turned and ran[8].

Further, the learning to fight was specific to the theater in which the soldier happened to be. In the Pacific theater, the combat soldiers learned how to kill Japanese and what to expect in their attacks and defense; in the European Theater, the soldiers learned how to kill Germans and Italians and what to expect from them. The two ways of killing and the expectations of combat actions needed were entirely different. As we will see in the next chapter, the specific way of fighting that the combat soldier learns is a rote reaction which provided immediate, unthinking, reactive response of the learned actions, in order to survive. Survival in one theater doesn't mean survival in another.

**

And then another high for the 102nd Division troops a few months after the European war ended, with the news of Hiroshima and Nagasaki hydrogen bomb destruction and finally the Japanese surrender and VJ-Day, the end of WWII, August 14, 1945. Finally to be going home! Going home. Going home. That's all most of us thought about. We were not the Regular Army; we were in for the

---

[8] That was probably attributed to the poor command section before General Patton took command in Africa, as described in Ken Burns' film, *The War* Episode Two

duration, and now that our job was done, going home was paramount.

The Point System

But, then, another low in the emotional roller coaster ride, as we were briefed on the point system for going home. We were told that the demobilization would be slow since **nine million American soldiers would have to be transported back home from all the WWII theaters of action, and thirteen million processed out of the Service.** No nation had faced a logistical nightmare like that before or since, and of course that couldn't be done overnight. The Point System was explained to us and at first we thought it sounded fair.

Then, we began to hear how the European Theater combat soldiers would be the last to go home because the point system favored non-combat and rear echelon troops, the Air Force and the Pacific theater troops. Three points were given for each battle star representing an official battle, such as Africa, or Italy or D-Day, or the push into Germany, all long duration offensives, not the often shorter ones the Pacific Theater experienced, island by island.

The Navy men and women and Pacific Theater veterans were given battle stars for each landing and clearing of islands, each one considered as a battle, so the active units got many battle stars, particularly the Navy units, and they were given priority to get home. Of course, the Navy had a built-in priority: they could just sail for home! The army had to wait for a turn with a limited number of transport ships.

Further, while each of us in the 102nd Infantry Division had three battle stars, or 9 points, the Air Force crews were given a battle star for each battle they *flew* over, plus their mission targets for each combat sortie, racking up scores and scores of points. More insulting, the ground crews that never left England were also given the same battle stars the air crews got!

In the European Theater, high points were also given for medals and awards, but the combat soldier didn't see many of them. It was far easier for the rear echelon brass to assign medals to the rear echelon do-gooders than to try to get medals to the combat troops fighting at the front lines, often being there for many months without rotation to the rear. How could there be ceremonial presentations at the front? Who would write up the combat soldiers' battle deeds? For the really tough battles where heroics happened, the only men left were tired guys who just wanted to rest.

Each battle and assault that was won was just another day of combat; sleep was all we wanted. Winning battles was never a sense of achievement for combat soldiers. If anything, trying not to think back on the attack made it a negative, not a positive, memory. Many of the higher medals were awarded posthumously, or were awarded years after the veteran had been discharged, too late to count in the point system.

The Heroes

I must make clear that I'm not saying that any medal recipient for combat heroism didn't earn them. Most of those men who received the higher medals more than earned them. I'm sure that the debriefings that would have had to happen were a digging process for the interviewers, trying to eek out information from men reluctant to talk about the circumstances. So the papers testifying to what the men did only stated what the awardees were willing to say when pressed, not the whole terrible story for each. Most combat veterans did not talk much about their experiences, not even to each other during lulls in combat or after VJ-Day in occupation. Any talk was about how to survive or how a recent effort would have been better if done differently.

I wasn't surprised in June 1994 when the media made much over the D-day 50th anniversary hullabaloo. Broadcasting from the beaches of Normandy, a young journalist was interviewing men who had been in the Normandy landing and fought on those beaches 50 years ago. At one point, the young journalist called one of the combat veterans a "hero on D-day" and the veteran got very agitated and angry and actually looked like he was going to take a swing at the journalist! Being back at the Normandy beaches undoubtedly was a huge emotional provocation and brought up not only the terrible memories which that veteran had, but also all the anger and PTSD behavior a veteran gets, as we will see throughout this book.

I've never met a combat veteran who thought of himself as being a hero. To combat veterans, the heroes were those who didn't come back, or men in other units which had much worse experiences for much longer periods of time, like the 1st Infantry Division or the Marines in the South Pacific island landings.

But then there also were the politically "manufactured" heroes, whereby the people at home were apparently thought to need a hero to go around giving talks, so that those people at home would feel less grim about the war, perhaps. "Heroes" were constructed and shipped around the country, giving patriotic talks to make people feel good. I suspect that media people developed the artificial, dishonest talks because it's doubtful that real heroes would have been willing to do that job, no matter how necessary.

If the above paragraph sounds like sour grapes, consider "Commando Kelly," a Medal of Honor recipient who did go on lecture tours around the country. I'm not questioning whether or not he rightfully earned the medal and the honor that went with it; it would be difficult to believe he didn't, and we'll never know what he really did. But the story he was telling around the country, probably concocted by a media storyteller, was pure nonsense. The military made the mistake of arranging my 102nd Division to hear his talk when we were at Ft. Dix, New Jersey, about to embark on the ship to go overseas. My Regiment was marched into a huge open amphitheater to hear him.

Kelly talked about burning out ten or twelve M1 infantry rifles, killing dozens of Germans one afternoon. And then he talked about banging the bases of 60mm mortar rounds on a window sill to eject the safety pin and then throwing them at the enemy Germans, killing dozens of them and saving the day. At first we were stunned at the utter impossibility[9] of all he was saying, and looked at each other to see if we were really hearing what we thought we were. Then we started to hoot and holler and when it looked like a riot was breaking out, we were quickly marched out of the amphitheater.

Although the 102nd Division had not passed maneuvers because the command sections were clumsy (or so we heard), we were well-trained infantrymen. Many of us were expert-certified in several weapons. We knew what our weapons could or couldn't do (see footnote [9], below). Further, we already had no patience with deceit or untruths, quick to anger and quick to aggress.

That story reflects the disconnect between combat veterans or those trained to be, and the rest of the army and civilian policy makers. Perhaps the storytellers thought we combat-trained soldiers would be somehow inspired by the story of patriotic heroism told by Kelly.

**

---

[9]   In training, we had used M1 rifles on attack-simulation ranges. We were grouped on both sides of an open range where hidden targets would jump up and you'd shoot an entire clip of rounds rapid fire as you ran across the range, then hand the M1 to a guy on the other side who would immediately start running back across the range, putting a new clip in the M1 and taking his turn to rapid fire at the targets, and so on, hundreds of men using the same M1 all day long. It could never be "burned out".

And we mortar-men had been shown in training that we couldn't accidentally actuate ejection of the safety pin by banging it on something, no matter how hard we tried, even two burley men trying. Besides, the 60mm mortars were not fragmentation rounds which could kill lots of men like a grenade would, but was concussive, to disorient and cause temporary confusion and blast casualties.

After the campaign up to the Rhine River and the 218 days on line, the 407th Regiment's 1st Battalion (including my B Company) was cited for a Presidential Citation, because we were the 9th Army's spearhead during that long and bloody offensive. That would have given us high points for going home, but at the time of the citation, the 102nd Division Headquarters declined the honor on the basis that if the whole Division didn't get the citation, no one would. The real reason was probably that the rear echelon brass wouldn't have gotten the Presidential Citation if it were given to only one battalion, and the rear echelon really liked getting medals. They were seriously fighting a war and took everything about the war as patriotic duty well executed, which would normally be their job. We combat soldiers weren't really seriously fighting a war –we were just trying to survive, and didn't care much either way about medals until the point system came out. Perhaps the 102nd Division rear echelon hoped the awards people would simply acquiesce to their demand for the entire 102nd Division to get the Presidential Citation, or maybe they didn't really care.

**

So the rear echelon with their extra medals got to go home first. When we finally went home after the occupation duties described below, Company B "old men" were designated to take home the 102nd Division and regimental flags since there weren't any rear echelon "old men" left to go home with us and do that duty. We were called "old men" simply to designate those who were original members of the unit; I was age 20 when we went home.

My squad (I was a sergeant, then) was filled with MPs from some other unit; arriving home was just another process because we didn't really know many of those we sailed with. We were no longer the 102nd Infantry Division, going home. The glory marching through ticker tape streaming down in New York City had been done long ago.

Occupation Of Germany

The reformed 102nd Infantry Division with new commanders and new upper echelon staff then became an Occupation Division for seven long months after VJ-Day before we could go home. There were near-mutinous actions at times, like at Deggendorf in southern Germany when the 84th Infantry Division (our sister Corps unit, also full of combat men waiting to go home) was relieved by the 102nd Division for occupation of that city.

During the changeover of occupation sites and command locations, a fight broke out in a bar between 84th and 102nd men, probably quite drunk which was our usual state during occupation. The confrontation flamed into a full-blown standoff where each Division had barricaded parts of the city, brought up its Canon Company's 50mm canons and set up mortars and machine guns. But for some quick-thinking command officers, we would have had a private war between the two Infantry Divisions. Some skilled officers should be proud that they managed to get the 84th units quickly out of the city before real bloodshed started.

Anger and violence were a part of occupation, since we were combat units that had been trained to be angry and violent. Once, when one of my buddies, Joe, was sitting on a billet stairway above the Sgt, Joe was throwing berries down on the Sgt's helmet to annoy him with the ping on the metal helmet. After a couple of growls, the Sgt spun around and threw his knife (we all carried) at Joe, hitting a major artery in his shoulder. We all laughed as we put a tourniquet on Joe and got him to the medic's station.

Similarly, once after VE-Day I had become a sort of local hero to a large displaced Polish group by capturing them, all armed and causing havoc in an area near our occupation city, with only one other G.I helping me. We single-handed brought them to the Polish community in the city we were occupying to be disarmed and billeted there so they could be sent home. The Polish community fronted on the same square where our billets were, and after the

28

"capture" of the armed Poles, they always waved to me when I was coming and going. Once, another G.I. and I were out in the courtyard, play-fencing, using some German officers sabers, and my buddy managed to jam his sword through my sword guard, through one finger and on into my right hand. I was so angry that I picked up my dropped sword with my left hand and started to hack at the G.I. with all my strength, really trying to kill him, while the Poles cheered me on. As I forced the G.I. back up a stairway into our billet, other G.I.s grabbed me and disarmed me, stopping the fight.

I had to go to the hospital at Gardelegen, Germany, but of course didn't get an additional Purple Heart! The doctor there mused that with all the accidental wounds coming through now, the casualty rate was almost as high as when we were fighting Germans.

Another remembered incident while at Deggendorf on the Rhine River, six or so of us got a riverboat fired up to sail home. Of course, we were pretty drunk, so while arguing about whether, in order to get home we should go north, downriver away from the Mediterranean, or south, toward the Mediterranean but upriver, the boat swung around and crashed into the pier. All we wanted was to go home.

We found more insanity at Le Havre, France, awaiting shipment across the Atlantic to home when the point system finally got around to us. The ship we'd first been told was to take us home was waiting for us in the harbor. We went to the pier to actually **see** it; we were really going home! But we then found that our ship had been politically abducted for French and German war brides to go to the states, and we'd have to wait another three weeks for the next troop ship.

More drunkenness and fighting. The platoons bunking together were mostly equal-pointers from many units, not the same guys we'd fought together with, so there was little camaraderie. Losing our ship wasn't betrayal; it was just the usual military SNAFU, right?

War's Desolation

Waiting at Le Havre, we were surrounded by the same depressing destruction we'd become used to. The entire harbor had been destroyed during its taking, and the piers being used for ship-loading were the portable docks that the Army Corps of Engineers had built. Realize that we shipped out of Le Havre almost two years after the D-Day bombing that crushed that city. The French people that lived there and tried to get some kind of order and survival going were living that way all that time and beyond.

While waiting for passage, we were bivouacked in 8-man tents and had a sort of PX to go to at least for cigarettes and booze. Homeless children, at least some probably orphans from the war, wandered through the rubble nearby us looking for food or coal. We gave them whatever we could take from the mess halls, but that didn't much help our spirits. We tried to keep the children from fighting with each other over the food we gave them, but they were just wild animals, no longer human beings. Once, several of us went out scrounging with them to dig up coal, a daily task the children did, taking the coal home each evening or selling it for food.

**

I didn't realize how immune we had become to our surroundings during combat until I recently saw a WWII picture (facing page) of a 102nd Division platoon moving through the rubble of the City of Linnich on the way to the Roer River. The picture was published in the 102nd Division monthly *Ozark Notes,* for March, 2008.

When I received that newsletter in the mail in 2008 and saw that picture, I was shocked at how different the rubble of a bombed out city seemed now, safely in America, 62 years later, but that is what we lived in during all our time in combat. What our Air Force and artillery didn't smash on our way into a town, and our tanks and mortars didn't bust while we were taking the town, the German artillery finished off while we were there under German fire before moving on. Some cities, like St. Lo in Normandy, France, did not have one single whole building standing –only some walls and rubble. Some church walls were still standing, but with no roof. The entire city was rubble.

MEMORIES

LINNICH—American Infantry marching through the ruins of Linnich towards the Roer. (February, 1945) Photo copied from the German book "RURFRONT" edited by Hans Kramp of Linnich, Germany, and printed in Geilenkirchen in 1982. Original photo probably provided by the 102ⁿᵈ Infantry Division Association.

Figure 3 – War's Desolation

I don't think American civilians can ever understand what it's like living in absolute destruction month in and month out or what it does to a person's psyche. The desolation was another form of death surrounding the combat soldier –the land around him was being killed, along with his buddies. The buildings, the roads, the trees, the farm animals, all dying. The combat veteran just insulated himself from seeing what the environment was – he floated through the days in a haze of denial as will be discussed in the next chapter.

Americans certainly can't relate to the life of people living in war-torn countries, not only desolation every-where, but no heat, water, or electricity, lost children and husbands, no food or jobs, no way to replace tattered clothing, or even to get a jacket or blanket to sleep warmly. For months. Or years, for most cities. When we fought our way into a German town, the people were gone, having moved farther into Germany ahead of us. But when they returned, what would they find? Homes and stores destroyed and looted, no utilities, old customers either missing or without money to buy anything, animals dead, businesses closed, neighbors vanished, streets so full of rubble that even a bicycle couldn't navigate. It was an entirely different world than we grew up in, one that deeply changed those who saw it and lived in it.

Realize that despite the horrors that I'm describing in this book, our war in the European Theater was more or less civilized, compared to the savageries of the Russian/German front. The Germans killed more civilians than military members, for a total loss by Russia of 25,000,000 people[10]; yes **25 Million!** Compare that to the American total of 486,000 military fatalities. **Russia sustained over two-fifths of the entire world's losses, 60 Million people, in WWII**.

**

Some of the disturbing anger we felt about going home, even if only subconscious, was knowing that many of the good B Company men killed in action would never go home. We sometimes talked about the 102nd Division fatalities buried at Maastricht, Holland, in the U.S. Army cemetery there. Long after the war, I recall news articles about bringing home the GI's remains if families wished to do so. I understand that most didn't. Some day I want to go there to Maastricht, find the B Company, 407th Regiment section, walk among the white crosses, and read out loud the names. I'm sure I would have heavy tears in my eyes, as I do now, thinking about it.

---

[10]   Data from internet: "Thank You Memorial.org"=

The Trip Home

The trip home across the Atlantic in the little Victory ship, a converted cargo ship, should have been joyful and full of cheerful memories, but we were too drunk, too crowded in uncompromising quarters and still fighting at the drop of a slur, to remember much pleasure. However, the trip back was faster than in the little Liberty cargo ship we went over on, in one of the biggest convoys pacing the slowest oil tanker in the convoy. And the Victory ship going home had all the good 'Big Band' music that we remembered and so longed for.

An example of our going home memories: one of our B Company "old men" who had served all 218 days of B Company's time on line without any break, was so screwed up that in his drunken stupor he tried to stand on the ship's railing to piss on the Statue of Liberty as we arrived in New York Harbor. We were busy trying to keep him from falling over-board and to not get pissed on ourselves in the blowing wind.

Discharge was a time where we were slowly starting to unbend and to realize we were truly home and out of the Army. But then, the logistics of thirteen million service men and women getting discharged was staggering, both to the veterans and to the civilian world. Because many millions of men and women had been discharged before we got home, we weren't all that welcome at the crowded veterans agencies, the colleges, the job markets, and the bars where civilians had already heard it all, over and over.

Sometimes the greeting at home wasn't ticker tape parades, particularly for the Vietnam veterans. We should never forget the cold rejection they were given. Although most of the country became a force against the Vietnam war and that was a large part of the war ending, most people seem to have convinced themselves that they really weren't against the veterans personally. However, all the Vietnam combat veterans I've talked to in various therapy groups over the years have agonizing feelings about coming home. They all felt the rejection and the isolation.

Many expressed how much they were more hurt by the people at home rejecting them than they were about the horrors they saw and did in the war. That they still bring up the issue of being so hurt even after all these years is testimony to their pain.

But that should be no surprise, because for most wars, the people who can't understand what war is, need to feel that they have been good to the veterans coming home from war, and so they convince themselves that the veteran is being taken care of, **in general**. In 1932 the WWI veterans couldn't find work or food or homes, and when they marched on Washington in 1932[11] to get the government to pay them money instead of Promissory "Certificates," 43,000 strong including some families, the country felt the veterans were going too far with needing help. The Army, commanded by Douglas MacArthur, was sent in, and "brutally suppressed" the veterans, killing some and wounding many and setting their squatters-built shacks on fire. Civilian onlookers cheered the army's actions.

**

I had a dismal experience upon coming home when I was waiting at New York's Grand Central Station for the train up the Hudson Valley to home. I, of course, was in uniform and trying to sleep on some steps in the crowded station. I was rolled (pick pocketed) by a guy who took off through the crowd and managed to evade my chase. My train tickets and money were in my wallet, but fortunately a USO booth there replaced my lost ticket, although I had to take the next train, another small delay in getting home.

James Jones, in his book, *WWII*, describes this period:
    And the civilian world went merrily on in its
    happy, dizzy whirl of prosperity in a booming
    economy. Articles appeared in women's magazines
    with titles like "Will He Be Changed?" *Good
    Housekeeping* said, "*After two or three weeks* [my

---

[11]   data from Wikipedia internet, article: **Bonus Army**

italics] he should be finished with talking, with oppressive remembering. If he still goes over the same stories, reveals the same emotions, you had best consult a psychiatrist. The condition is neurotic." *House Beautiful* recommended that "home must be the greatest rehabilitation center of them all" and showed an (elegant) apartment fixed up for some homecoming General. *Ladies' Home Journal* asked, in1945, "Has your husband come home to the right woman?

The answer, of course, was no. How could any woman be the right woman for a man who had just spent one year or two years as essentially a dead man, waiting, anticipating having his head blown off or his guts torn out? (And watching it happen to one after another of his buddies.)....

Instead of talking about it, most men didn't talk about it. It was not that they didn't want to talk about it, it was that when they did, nobody understood it. It was such a different way of living, and of looking at life, even, that there was no common ground for communication in it (Jones 248).

And Ken Burns' documentary ***The War***, depicted this disconnect as narrated by a veteran, describing a letter he had started writing to his sweetheart, aching to try to bring her into his world of death, to share his horror, an important part of his life, with an important someone who loved him. But he never could send it and carried it with him until after the war.

As he read it:
"I am not the same person you said goodbye to… No one could go though this and not change. We are all casualties. In the meantime, we just go on. Someway, somehow, this will all have an ending." (Episode Five)

In the next chapter of this book I describe how irreconcilable the two worlds of the returning combat veteran are, and how difficult they are to live in and to communicate between, the world of combat reality and horror, and the civilian world of safety and friends. The combat veteran had the added agony of needing so badly to share his feelings and thoughts with someone who cared about him or her, but there was no way to bring someone not a combat veteran into the world of horror and fear that he or she lived every day.

Benefits for the Returning Veteran

To assist the veteran and family with the difficult readjustment needed upon arriving home and discharged, both the Federal Government's Veterans Administration and the States' veterans agencies have processed and continue to process many benefits for the returnees. Congress and local Legislatures have quickly pushed through many benefits after each war.

This book's *Appendix A, Veterans Benefits* is a listing (page 81) of benefits currently available to veterans, at both the Federal and State levels. *Appendix B, How One Goes About Getting Veterans Benefits* is a listing (page 83) of the many agencies in counties, cities, local areas, veterans' organizations, and military unit organizations which are dedicated to help veterans and families get the benefits listed in Appendix A. Much help is there; all you have to do is ask for it, although that may be hard for some veterans, who find it difficult to ask for help, as we will see in the next two chapters.

One of the most helpful benefits for veterans is the work of the Veteran's Administration Hospitals located in most medium and all large cities. They have been providing extensive medical help and prescription medicine to any combat veteran they have determined has service connected problems at no cost to the veteran.

36

Chapter III
WHAT IS PTSD?

A Brief History[12]

In ancient Rome when a Centurion became a combat veteran, he was then forbidden to ever again enter the city of Rome, even if that was his home and birthplace. Why? Because when combat veterans come home, they are angry, loud, unclean; they drink and do drugs or obsess with pornography, abuse women, get in fights, use endless vulgar profanity, dress bizarrely and grossly, answer to their own law, not ours, like to gang together, and are not much wanted in polite society.

Most nations through history have sent combat veterans to rule conquered countries, or stoned and exiled them if they had lost the wars, or let them drift into areas of cities and towns that were ghettos of crime, violence, and poverty and made them feel unwelcome elsewhere in the cites.

After the Revolutionary War, our country sent combat veterans to quell the frontier and take any land he could hold. After the Indian Wars and the Civil War, combat veterans were sent to remote areas out West to homestead. The Indians were also combat veterans--all men, women and children--were given useless lands and kept there by force in reservations.

After WWI, we called the combat veterans' afflictions "Shell Shock" and shunned the veterans until they died or had drunken accidents, and again giving them home-steading rights in a few remaining desolate or isolated places, hoping they would be too far away from Washington, D.C. to participate in the many demonstrations for government help that went on. After WWII we

---

[12]   Most of the information in this Section, *A Brief History*, was from the California Veterans Center counselors at 1111 Howe Avenue, Sacramento, when I received counseling there in 1991.

called it "Battle Fatigue," and –guess what—there weren't any places left to send the combat veterans, so we put them in jail when they were unsociable. At the time I was attending the Combat Group Therapy sessions in 1991 at the California Veterans Center, a significant percentage of the inmates in California prisons were combat veterans from WWII, Korea, and Vietnam, but were receiving no VA or Cal State PTSD treatment benefits as I was, because the prison authorities didn't allow it. (See "A Call for Action," page 79.)

After the Korean War, the affliction was called "Battle Neurosis." Following the Vietnam War it was finally called PTSD–Post Traumatic Stress Disorder–and since then it has been considered a treatable disease by social workers and therapists. For the average civilian, however, veterans were people you didn't want to live next to or to associate with socially. By any name, PTSD has been an affliction which combat veterans coming home from any war have had to somehow conquer or be outcasts.

Sadly, however, some of the health service field's professionals and media[13] still don't consider PTSD a physical disease, but instead a psychological disease, for which treatment benefits for veterans are subdued. Although many of the VA counselors **do** consider PTSD as physical damage as we will see in this chapter, (a few sections below: "**Medical Science**"), this area of explosion blast concussive damage still seems to be in the "dark ages" of trying to call the affliction something for which Society doesn't have to be responsible.

The article on *Brain Injuries From Blasts*[13] indicates that this new area of understanding of combat affect is

---

[13]   Reference New York Times newspaper Health Section Tuesday, June 23, 2009, page D6: "*A Chance...Brain Injuries...,*" where, talking about brain injuries from blasts, " Disability benefits for veterans can vary depending on whether an injury is considered psychological or physical. And veterans with P.T.S.D. alone do not receive the Purple Heart, the medal given to soldiers wounded or killed in enemy action, because it is not a physical wound."

becoming studied and applied to combat disability as real, physical wounding. It is slowly being explored as one of the aspects of PTSD. Why it is becoming suddenly a studied aspect is difficult for me to fathom, knowing that PTSD was called "shell shock" as long ago as World War I!

All combat veterans returning from foreign wars experienced at least some of the insanity of war, even if not wholly in harm's way. In the Vietnam War and the Iraq and Afghanistan Wars, most service men and women, even non-combat members in support units, were in harms way most of the time. In those more recent wars, there wasn't much safe haven anywhere, no safe "rear" like the WWII and Korean Wars had or the R & R (rehabilitation and recreation) areas of the South Pacific in WWII. Those wars proceeded according to the logistics of the time, when major countries fielded large armies that could be supported only with controlled rear areas and protected routes of supply. Today's wars are entirely different. Most of today's combat veterans and many of the support veterans have seen enough horrific insanity that they all have PTSD, some more than others.

What Is PTSD?

The words "Post Traumatic Stress Disorder" don't tell you much, but there are many books written about it. "Psychologists classify PTSD as an anxiety disorder characterized by frequent re-experiencing of a traumatic event (or many events), emotional numbing, avoidance of reminders of the upsetting event, and excessive vigilance."[14]

It is what happens to your psyche after a traumatic experience and when grieving can't be done. Of course, many people and including children who have never been in combat have PTSD because of horrific experiences they unfortunately have had in their childhood and which didn't

---

[14]   Excerpt from "*Science News*" weekly news magazine of science, vol.173, No. 1, "Damage Control "

get properly grieved due to the many reasons in our culture for denying real healing. And, unfortunately, many children throughout the world have been and are being exposed to the ravages of war, although not many of our American children.

Everyone experiences some traumas in their life, often not just one horrific experience, but those usually separated by years such as a child's death or a bad car accident. For most traumas, a person can go through the grieving process for each, and resolve the trauma in some sense, so that it is put to rest. Combat traumas, however, go on nearly every day, sometimes in multiples for hours and continuing for months or years, and with intensity you'd think would drive one crazy at the first experience alone.

The worst war trauma intensity I've heard about was in a WWII documentary video about the Marines landing on the beaches of Saipan Island in the South Pacific Mariana Islands. The first waves of Marines were pinned down under constant fire on the beaches for 21 hours. Imagine, 21 hours of incoming artillery, mortars, and withering machine gun and rifle fire, and the naval guns' barrages going overhead in a constant din, trying to check the Japanese counterattacks. Nowhere to go; nowhere to hide; lying out in the open; 21 hours of endless explosions and death all around you. Those men were all heroes, both those who lived and those who died.

Further, for the more recent Middle Eastern wars (and actually, all wars), many combat veterans went into the military service to try and get away from the PTSD stress in their childhood and home life. The youth grown up after a disadvantaged childhood finds purpose and self-worth as a military person. But those veterans who went into the service already handicapped with home-grown PTSD must experience a double whammy if they are sent into combat. They not only must experience continuing traumatic experiences, but also at a huge upgrade in the intensity and frequency of them. Therapists teach that a child growing up with unresolved PTSD is more vulnerable to getting

PTSD from other traumas later in his or her life, and will get deeper and more anguished PTSD than those with functional upbringing who experience the same level of trauma.

No Time For Grieving

Each traumatic experience (non war related) will automatically be processed if other people support the grieving person in the process, and when the grieving person is given time to resolve his or her loss, for each trauma. Bereavement counseling teaches that time needed for grieving a given trauma, even intense traumatic grief, is six to twelve months.

But of course there is no standing aside in combat to deal with the emotions as each day's traumas happen, not then, not weeks later, not years later --for some, not ever, without counseling help. The counselors stress that any trauma we have in life must be grieved if we are to retain our emotional sanity. Each trauma must be grieved, in real time, through all the separate steps of grieving: denial, anger, negotiation (with self or God), despair or depression, and finally acceptance.

Suppressing and denying grief for the horrors going on in combat is what creates the stress that takes enormous energy to keep going when there is no opportunity, no time or place, for grieving. After a veteran comes home, the same enormous energy is needed to try to act "normal" and to keep a relationship on some real level, or to get through school or to do a responsible job at work every day, but social function and productivity at work and at school are often significantly compromised.

**

James Jones' *WWII* described many examples of the kinds of utterly brutal, ugly, even criminal experiences a combat veteran was supposed to forget, but can't. His book expresses what really happens in combat, not the glorified version which many books written about wars create. And Ken Burns' film *The War*, shows much footage of

documentary film taken during the war, showing some of
the shocking atrocities committed and our brutal responses
to them, sometimes criminal.

I remember when I was wounded and trucked back
through Liege, Belgium, then by train to Paris on the way
to England by air, I was of course a hurting, scared
nineteen year old, feeling betrayed, and barely conscious
that only days before, I had been trying to kill Germans.

When I arrived by ambulance at the hospital in Paris
from the train station and was being pulled out of the
ambulance, I was shocked into fury to see that the men
carrying the stretcher were German soldiers, still in their
uniforms! Without any conscious thought, I lunged at the
one at my head, upsetting the stretcher and I fell off it to the
ground. The German prisoner I had tried to get at was
crying as he was led away, while the GIs soothed me and
got me quieted down. I was told later that the German
prisoners knew that if they dropped a wounded American
GI, they would be taken out behind the hospital and shot. I
felt better back in 1944, knowing that.

Medical Science

Scientists have identified portions of the brain that have
become different due to the numbing stress that combat
veterans go through. It is real; it is physical and sometimes
beyond human ability to cope, without therapy. Scientists[15]
are finding that PTSD involves over-activity of the
amygdala structure in the brain which, in cases of PTSD, is
affected by the suppression of the fear response by the
prefrontal cortex, the area of the brain where sitting on your
emotions shuts down the immune system.

PTSD also creates a blockage in the prefrontal lobe of
the brain which normally assigns common everyday
memories a storage place in other parts of the brain, under
the general label of "PAST," and they stay where they are

---

[15]   Excerpt from *"Science News"* weekly news magazine of science,
vol.173, No. 1, "Damage Control "

assigned until you retrieve them at will, or not at all if you choose not to go there.

But the horrific experiences that the combat veteran has survived get blocked from the storage assignment process and don't get stored or labeled "PAST." They wander around the brain like ghosts haunting a house, unable to get to the resting place they should have. The memories resurface time after time, getting unasked and unwanted entrance into consciousness when some other associated "trigger" sets the memory into motion again. The combat veteran relives the war experiences over and over and over, in distressing nightmares and as daytime flashbacks of intrusive thoughts or feelings. The visitations are real and intense, as though the veteran is back in combat, hearing and seeing the horrific occurrences in real time, again and again.

It's like an experience you may have had, common to most of us, where a little tuft of debris, perhaps, lying on the floor in peripheral vision and perhaps moving a little in a gust of wind, suddenly is perceived as a black widow spider about to climb up your leg. You startle; your heart races; you swing into a defensive position and focus intensely with every sense alerted to the threat; and then laugh when you realize it wasn't the spider you thought you saw, just some debris on the floor. However, what you thought you saw was **very real** for an instant, and your reactions and emotions were exactly what they would have been if the spider had in fact been real.

PTSD's unwanted flashbacks and nightmares cause the same reactions and emotions in the combat veteran as if they were in fact real, because, like the spider, they **are** real to the veteran having them. He or she is, for an instant, back in combat reliving all the sounds and images that were real and the huge emotions that they generate. They happen over and over until the veteran thinks he or she is going crazy.

Characteristics of PTSD

Veterans from any war have a hard time sleeping, have nightmares and wake often. Sometimes the nightmare is so intense that the veteran shouts out in his sleep, usually unintelligible yells, and in thrashing around might even fall out of bed. It is difficult for a spouse to be civil when woken out of a deep sleep with the veteran yelling or thrashing about in a war nightmare. If the spouse puts a hand on him to wake him out of his nightmare, the spouse could get seriously attacked before the veteran fully wakes out of the intense nightmare. The veteran feels horribly guilty and the spouse if attacked will have trouble feeling love for a while. Some of the saddest affects of WWII were the thousands upon thousands of marriages that ended in divorce because of the combat veteran's PTSD affliction.

Combat veterans not only have recurring war nightmares, but also have many recurring nightmares like any non-veteran with home or work bred PTSD might have. Common themes are being chased by someone or something which may be a war something (tank or an enemy soldier with a weapon while you have none, or even artillery fire chasing you across a landscape), or just an endless chase by someone or animal or something that goes on for seeming hours, until waking up in a sweat just as you are about to be caught or killed. Another common theme is bounding great distances at great heights and then coming down fearfully fast from the great heights and crashing on the rocks or roads, but then finding you've bounced back up and are careening again on a flight toward another fearful crash, on and on all night. Although the crashes miraculously don't hurt you, each fall to the ground is horrifying, and the dreamer might wake up at that point.

The spouse or children coming into the house might be greeted by a blood curdling "primal scream" from the veteran who thought he was alone, remembering one of the many horrors, and letting out the agony of the memory with the scream. Veterans find that doing the primal scream helps blunt the intensity of a flashback, perhaps because the scream is what he or she did in combat, so it seems to make the memory not a fantasy, but part of reality. Conventional wisdom says that primal screaming is an expression of emotions (anger or rage) and therefore helps keep your immune system healthy by letting emotions out, whereas suppressing emotions shuts down one's immune system, which is not good for one's health. In any case, therapists teach that primal screaming doesn't make the flashbacks happen any less often or go away.

A common feeling among most soldiers in combat is that when the traumas are happening, the soldier feels helpless to control anything, and not able to change or stop anything from happening. It is an anxiety very much like being tied up with no way to get freed from the bonds. It's a helplessness that is a part of the trauma and the resulting PTSD. Not surprisingly, another nightmare theme that recurs often for veterans is combat action where the Veteran can't make something stop, although he or she feels, in the dream, that "I could have made a difference if only I had done something….," or in the dream the veteran is simply a bystander watching some horror happening.

Combat veterans, if incarcerated, are more prone to express their anger and violence than other prisoners, because the incarceration puts them back into the helpless realm of combat. Any situation which makes the veteran feel helplessness and loss of control, will recreate combat anger and anxiety, sometimes subtly but sometimes with reactivation of a flashback as though triggered by a memory.

Triggers

The veterans feel guilty at not being able, as many veterans have been told, "to put the war behind you and just get back into the good life." But they can't seem to help reacting to the triggering of combat memories by war kinds of sounds: a car backfiring, or firecrackers on the 4th of July sounding like a machine-gun firefight, or even the sound of helicopters going overhead, for Vietnam and Middle East

War veterans. Once, shortly after I was discharged and was home, I was standing on the front porch, which was about eight steps high, talking to my Dad when a car backfired passing nearby. Without any conscious thought, I dove over the porch railing and landed in my Dad's prized rose garden!

I can laugh about that now, but I felt sheepish and puzzled then, and guilty to have behaved "foolishly." Even though my Dad understood, I didn't--and that's what a veteran feels all the time--his reality is something different than what everyone else seems to be thinking. It's not just puzzling to the veteran; it's scary! The veteran comes home from a scary existence for many months or even years of combat in a far-away land where there is no connection to the life the veteran knew before enlistment.

Even though the veteran is constantly in harms way in a war zone, everything going on follows predictable, familiar patterns. Even in ambush, the combat soldiers know exactly what is happening –it's happened before and will happen again. The soldiers know exactly what the Sgt is going to yell at them, and exactly what must be done and with exactly the same urgency and skill. Often the Sgt sees that he doesn't have to say anything; the soldiers know what to do. As pointed out in Chapter I, the Pacific and the European Theaters of War were entirely different kinds of battle and produced entirely different survival learning. However, the process of learning and then applying those survival skills quickly and unemotionally was the same for all combat veterans. The veteran follows his stateside training and his combat survival learning, and it seems to work for those who stay alive.

Now at home, everything is still scary but in a different way, with unfathomable mind-games that everyone is playing. How do I get out of this, the veteran keeps thinking. Some feel more comfortable going back for another hitch in the service, even if it means combat, than trying to survive in the scary, unpredictable civilian world.

**

Each veteran has different memory triggers that give him or her flash-backs to war experiences. Veterans often cry or get angry or depressed at what seems to others as no real issue; a veteran's behavior sometimes doesn't seem to make any sense to non-veterans. It could be anything--a sound, a flash of light or lightening, a cooking smell or even a whiff of sweat or a sulfur-like smell as from an explosion, a person's voice or a particular situation such as seeing a child hurt or hungry, a moving shadow seen in peripheral vision or the way a person in the distance is standing, part of a song, or even newly-planted fruit trees with white-painted trunks, lined up like crosses in a military cemetery in Holland.

I remember a work-mate in the 1950's who had flown twenty-two B-24 bombing missions over Germany, often coming home to English airfields with the plane full of holes and wounded crewmen. He had nervous flashbacks whenever a large plane (propeller driven, then) flew overhead (our workplace was under an airport's final approach pattern). He admitted that he was adamant about never, ever, setting foot again in an airplane.

Those triggers are sometimes so subtle that the veteran seems to just have intrusive thoughts or feelings, not being conscious that an unobtrusive external trigger may have set off the remembrance. There also are **internal** triggers, from dates that are anniversaries of horrific actions the veteran went through, or even "war holidays" like Veterans Day, Memorial Day, 4th of July, or VE Day, VJ Day, and D-Day. For those subconscious triggers, the nightmares or flashbacks seem unprovoked and thereby even scarier.

The shock value of differing triggers has different intensities. Smells that trigger flashbacks are the most powerful, because our brains have been wired from early animal evolution to send smells directly to the Limbic System of the brain (including the Olfactory Bulb near the central, lower, brain stem), where there is direct connection to the fight or flight wiring of the brain for immediate action similar to the direct wiring of touching a hot stove top. All other sensations go through the prefrontal lobe and are processed/analyzed first, and are not directly wired for immediate survival response.

Numbing Out

The combat veteran learned to survive combat by numbing out, which is what the PTSD counselors call it. To stay sane in combat's continuing, intense traumas, the combat soldier's brain does a peculiar trick: emotional memory and emotional processing goes unconscious, so he or she can continue, undistracted, in the business of combat. Normally, every soldier would need to grieve every terrible thing that happens, as we discussed earlier in this chapter. The soldier needs to grieve every buddy he or she sees getting killed or maimed, and the enormous guilt the veteran feels, not being able to stop to help or comfort the buddy. He or she needs to grieve every trauma and loss, every tired day with no hope of getting home alive, and *grieved right there in real time,* or at least to hold his or her dying buddy.

But--and it's a **huge BUT**--there is no time or place or pause for grieving when the combat is happening, with its intense rush to do the right thing and not let anything distract you. So after a while, the combat soldier goes into an emotional cave, feeling nothing, which is how he or she survives. The PTSD counselors teach that without that numbing out, the soldier would certainly go crazy. Some did, even experienced combat veterans like in the movie *The Big Red One,* when the American soldier at Dachau is shooting an already dead German soldier who had hidden in a crematory oven, shooting him over and over and over, until the GI's rifle, empty, is just clicking, over and over, as he keeps pulling the trigger. And his sergeant, knowing, quietly walks away.

A human being can take only so much, everyone differing in the threshold of physical or emotional pain that he or she can take before cracking. There is no formula that the combat soldier can use to ultimately avoid that breaking point if pushed that far.

In addition to not having time for grieving, the numbing out also helped the veteran stay alive by being calm and collected and solving whatever problem was on the horizon. Good for combat or other crises in life, but not good for relationships, to be calm and collected and detached, no matter what is happening.

That is what he or she learned to survive in a war, to solve the problem, stay calm, keep going, take action, solve the problem quickly, no matter what else is happening, solve the problem, keep going, stay calm, take action, stay calm............ That translates into a seemingly deliberate aggressiveness, born of beach landings, river crossings, large-scale attacks by either us or the enemy. It's an emotionless "get out of my way" mentality, but it kept us alive.

While combat driven numbing out is an alertness-enhancing behavior, where the veteran is calm and collected and thinking with great nimbleness, speed, and accuracy, it is not depression. It is as though part of the brain, the emotion-controls and expressiveness functions, become "unconscious." Although the veteran will vividly remember every visual, sound, and smell experience that happened in combat, he or she usually can't remember emotions experienced while in the thick of combat, even though he or she may have been screaming with rage, face contorted with fury, eyes blazing with hatred and fear. His or her buddies might reiterate that scene, but the veteran will just have blank confusion, no memory of what he or she was feeling.

A mental mechanism seems to have been inserted by trauma somewhere in the brain circuitry, whereby the veteran will automatically and unconsciously turn off emotional consciousness when trauma induces his or her combat mode of anger and attack. The combat veteran can't survive if emotions distract him or her from the necessary focus on combat and staying alive.

A peculiar anomaly seems to happen concerning those feelings. Although the combat veteran numbs out feelings during combat and even in discussions between combat episodes, the PTSD flashbacks and nightmares are rich with remembered feelings and emotions that are a real part of the terror and horror of PTSD.

Similarly, in the thick of battle, so much noise is going on, with artillery explosions and machine gun fire, that the memory of noise is selective. I can still hear the particular sound of a German rapid-firing machine gun, one of the most deadly ever invented, or the sound of artillery rounds coming in, or ours going overhead. During combat, we needed to process those sounds and react to them, to survive. But memories of combat scenes seem to be eerily quiet, no sound. Perhaps the mind closes off that memory, also. Sudden explosions that surprise one seem to be remembered, but on-going sounds in an attack don't.

One such memory I can recall is during the attack on Welz when I was trying to take a picture of the 81mm mortar rounds (friendly fire) marching toward me through the woods, each round kicking up a cloud of dirt and debris. The memory has no sound; it even seems to be peaceful and serene. Then one mortar round landed right in front of me, its explosion blowing me several feet into a haystack, luckily unharmed (who was my angel that time?). Even the memory of the explosion doesn't have any sound.

Suppressing Feelings

In suppressing feelings instead of grieving, the combat veteran carries the burden of grief with a huge sense of loss, of guilt, and of shame. That burden isn't left back in the combat zone when the veteran returns home, but is carried into his or her life as a burden that colors everything the veteran does. After returning home, the veteran still has that huge burden of un-grieved traumas, and as memories surface of a given trauma, the veteran gets stuck in a part of the grieving process for each.

All steps of grieving but the last step, "acceptance," are really ways to **NOT** grieve. Denial, anger, negotiation, and despair are subconscious defense mechanisms to avoid really letting the deep hurt come into consciousness until it can be handled and accepted.

As one of the flashbacks or nightmares happen, the veteran will start through the grieving process, but get stuck before he or she gets to the "acceptance" part. The veteran has become so conditioned to not talk about the horrors, to try to suppress them and to put them behind and get on with life, that the same point in the grieving process is reached with each flashback: anger or denial or despair, which becomes part of the PTSD instead of resolving to acceptance. Only therapy or perhaps a loved one's special, understanding and TLC can help the veteran get past the stuck place where he or she may be.

Further, although there may not be horrendous traumas in the life of the combat veteran after returning home, there are situations that provoke the same response he or she had in combat. When threatened with any situation in a relationship or job or car problem (or even a tuft of debris that appears to be a black widow spider) which prompts the "fight or flight" response, the veteran numbs out and avoids. With the training we discussed in Chapter I where only fight, never flight, is a fixed part of the veteran's psyche, but now at home where fighting isn't socially acceptable, avoidance is the only option left.

Most veterans feel that they somehow do not fit in, in almost all areas of their lives now --the wife, the children, the house, the job, the boss, old civilian friends, the social parties, etc.--and when threatened, the veteran numbs out with drinking, with drugs, with gambling, with depression.

Depression

Depression is the most common symptom of PTSD behavior after discharge. The PTSD-generated depression is like dysfunction-driven depression, whereby the veteran just stops functioning, does nothing, says nothing, doesn't even seem to be thinking anything. He or she may sit for hours, staring at the wall or out the window and will show great annoyance at being pulled out of his or her stupor. Again, the emotional part of the brain seems to have gone unconscious like in numbing out, but in addition, the action part of the brain also seems to go unconscious. It's as though there is an infection in the brain, where the numbing out takes over not just the emotional consciousness, but other parts of the brain until there is little consciousness left.

I think combat veteran depression is like a dog barking when left alone and cannot either fight or flee when there is an uncomfortable sound. Experts say that a dog thusly barks simply because it doesn't know what to do. Combat veteran depression may be the same thing, where training has sealed off flight as an option, and fighting is no longer an option in the civilized world of home, so what is the veteran to do?

Suicide rates of combat veterans are as high as for any kind of depression. Many combat veterans at one time or another have tried to kill themselves while in a state of depression and inactivity and the agony of re-experiencing combat traumas over and over, or their civilian lives seem to be endless social and emotional problems. If they don't succeed with suicide, it's just one more puzzling trauma and new guilt to add to the PTSD.

Terror of the World of Combat

Part of the damage of PTSD is the terror of a world that is disordered and unpredictable from moment to moment, for most of the veteran's time in combat. Alva Noë, in his book *Out of Our Heads*[16], makes a good case about mind-world relations for seeing people not just as a biology separate from the environment and other beings and animals. He offers that we are inseparable from our webbing and connections with our physical, social, and

---

[16] *Out of Our Heads*, Chapter 6, "The Grand Illusion", page 141

cultural environment, and even one's daily activities that are interwoven to the point that they are not just aspects of the person; they **are** the person. Attempting to, or being forced to, make a transition to a wholly different world can be devastating.

More importantly, Alva Noë points out that we not only are culturally webbed into our environing world and it's constancies, but we **depend on** seeing the world around us "as a causally, spatially, temporarily well-ordered, regular, and predictable place," if we are to retain our sanity. In the combat world, none of those descriptions apply, and the veteran has a constant terror of what is going to happen next. Not only is it terrible stress and test of his or her sanity, but since it is forced on the veteran with an all too brief training and then the rush into combat, it becomes a real and unyielding terror, just to stay alive from day to day.

Regardless of how far the veteran becomes reacultured into the world of combat, it always retains its terror. Regardless of how inured the veteran gets to the unpredictability and disorder and desolation of his environment, he cannot forget that other world of home and order and friendliness which he so yearns for.

The final trick of war is that the combat soldier can see patterns of predictability in the chaos and insanity of combat, and can learn to survive by responding to those predictable outcomes and apply his learned capability in combat to new, unpredictable situations, as though it were the same old combat –and it works, for those who survive.

Two Worlds

It is important here to recall what we learned in Chapter I about what has been happening to the veteran while in combat. The next Chapter discusses many of the specific changes that have happened to combat veterans that make the veteran what he or she is, most usually brought on by accumulated PTSD. But another generalized factor is at play; I like to call it "Two Worlds." The combat veteran has been living in an entirely different world than the safe, friendly, orderly world back home; he or she has been living in a horrifying, emotionless aloneness and desolation and constant unpredictability that created an entirely different world. It is not new psychology theory that the affect on a person of living in an entirely different world changes a person. Even when a veteran is put in harms way for a brief few months, he or she can be changed significantly because of the intensity and horror experienced; those who suffer the experiences of war for several years are certainly wholly changed.

Of course, even while the veteran becomes acculturated to the world of combat, he or she clings to the memories of his or her old world of the familiar family. The agony of trying to keep one world alive while living in another is a part of the agony of PTSD. The veteran needs connection to his family and loved ones, to have any sense of sanity when returning home or even while in combat. In WWII, letters were the only connection we had, but in this century's wars, with electronic technology, soldiers can talk often to those back home, or send and receive recordings. In any case, keeping the connection alive is critical for the well-being of the veteran.

And yet how can the veteran share the horrors of war with a loved one? Just as the veteran in Ken Burns' film *The War* agonized over writing about his horrors (reference pages 35 and 36 in the previous chapter of this book), the veteran wants so desperately to not let go of the other world of home. He subconsciously knows that he is slowly slipping into a new world and feels he is losing the other world. The farther he or she slips into the world of war, the harder it will be for the veteran to get back into the world of home and orderliness, because every day in combat changes the veteran a little bit more, until there is little left of the old person.

As authors have written, when a person is immersed in a wholly different culture than he or she was used to, the acculturation creates a new person living in a different world. Alva Noë[17] expresses the factor that must be understood here: that the world we live in, with all our social webbing, emotional connections and on-going experiences is who we are. We can't just slide through a looking glass and be someone else in a different world. The two worlds, that of the home and childhood, and that of the horrifics of combat, are not just two cultures that the veteran can be expected to easily re-acculturate in and out of. They are two worlds that require **a different person** in each. A person simply cannot be two people at the same time, trying to live in both worlds or communicating with someone else in one while the veteran's reality is based in the other.

Likewise, the people at home, when the veteran returns, find it just as difficult to communicate with a person they don't know but think they do. But the veteran has been changed into a different person, with different needs, different expectations, and entirely changed reactions and focuses. In time, with therapy, the veteran can gradually re-acculturate from the combat world to the civilian world, but, as noted in the next chapter, that transition is full of quagmires and surprises, and can't happen without a lot of time and TLC given to the veteran. That TLC is as difficult for the loved one to give as it is for the veteran to gain from it, as we will see in the next chapter.

VA Counseling

Some veterans have lived with PTSD for fifty or sixty years (WWII combat veterans), at great cost to their relationships, to their productivity, to their financial well being, and especially to their health. Today, PTSD is recognized as a curable disease, with therapy. But PTSD is rarely "cured" entirely. At least, therapy can provide a reduction in occurrence of flashbacks and nightmares and in the intensity and the debilitating effect of them. In almost all cases, the veteran can be given back meaning

---

[17] *"Out of Our Heads"* Alva Noë

55

and control of his or her life, but the triggers are so deeply buried that sooner or later, PTSD attacks will recur. As so clearly expressed in Ken Burns' poignant and very real documentary film, *The War,* a WWII veteran is recounting many years after the war, how a combat veteran just has to deal with PTSD as a part of his life after combat, just as he had to deal with horror of the combat world back then. The combat veteran, who could accept that he or she was going to die but kept on fighting and functioning effectively, will find the flexibility to cope with the burden of PTSD.

Techniques can be learned to avoid triggers such as war movies, certain topics and places. The veteran can learn that certain of life's social functions like going to funerals and cemeteries can be expected to provoke flashbacks or nightmares, and that knowledge can help minimize the affect of the trigger. The veteran can learn how to be conscious of the sequence starting a PTSD attack and consciously change his or her response. That is, the veteran can learn to do consciousness guarding in order to recognize a trigger when it starts and learn how then to minimize flashback.

PTSD-resolving techniques also include writing letters to buddies killed, to civilians (in general) killed by the veteran, where the veteran is stuck in the feelings of guilt and remorse over what he or she did or didn't do, whether or not the veteran is to blame. Letter writing or journaling is particularly effective to get the veteran unstuck in the grieving process for those memories that keep coming back, unresolved.

**

The Veterans Administration (VA) has been providing extensive help to any combat veteran they have determined has PTSD, at no cost to the veteran. The therapists use an array of many techniques to help the veteran minimize the combat memory recurrences. For example: Rapid Eye Movement (REM) techniques to replace or associate the triggers that cause flashbacks or nightmares with newly constructed good memories. They use both one-on-one psychotherapy and group therapy. They provide gentle drugs to help relieve depression. They have many specialized programs like Anger Management Workshops. The VA is a large part of the benefits discussed at the end of the last Chapter and shown in the Appendices A and B.

The Marriage Relationship

I want to speak a special word about marriages which are so sorely tested by one spouse going off to war. It is terribly sad that all the help for veteran and family is predicated on the marriage partners realizing the issues and quagmires that they have been going through, but that often doesn't happen until so much hurt and distancing has occurred that the partners lose a sense that there is any reason to try to solve the issues. There are two factors here that are overlooked.

Firstly, the two worlds and the changed person that has come home from combat, those issues we've been discussing throughout this Chapter, are not always addressed as they have been here, noting that there is a huge communication difference to be worked out before a couple can even start relating to each other again. Understanding that there is such a chasm of difference between how the two think differently, react differently, and prioritize everything in their lives differently, is needed before the decision to get help from VA counselors can be even considered, let alone the decision accepted and the help gotten. And once the decision is made, it takes great dedication and caring in order to keep the focus on getting the help that is needed. Much distress and discouragement will be a part of getting therapy, and must be jealously guarded from derailing or sabotaging the help that can be had.

Secondly, most people aren't really aware how they interface with each other through learned, associated responses which are sometimes called "tapes" in their heads. As with all human behavior, it is natural to reduce all actions to the least amount of energy expenditure for any area of life. And being a partner is no different. We get into almost "ritualized" behavior toward others which are significantly in our lives. We say things the same way we have before, reducing the communication to saying the same things, showing the same emotions, and using the same emotional cues and behaviors that worked before, often lapsing even into sub-conscious analysis and resultant behavior for a given, usual, interaction. Everyone is familiar with the "argument tapes" we replay when we have a disagreement and discussion about it, saying the same things (sometimes hurting, sometimes nurturing) in the same situations that were said before.

When the veteran comes home, the tapes don't work any more because he or she is a different person. The usual reaction of the spouse is to feel disconnected and not loved any more, since the tapes are a reflection of the TLC and kinds of caring that were there before the veteran went away to war.

And it follows that not only is conscious communication affected by all the factors we've been discussing in this chapter about PTSD, but also the unconscious "tapes" which don't work anymore and the discouragement which that creates. They are tried over and over again with disappointment over and over again, gradually building a resentment "case for our side" which is hard let go, even with counseling.

A couple seriously seeking help to rebuild the relationship must realize the reality of this book, and particularly the next Chapter dealing with what the combat veteran is like. Like it or not, the combat veteran is a product of what politicians decided was an O.K. ancillary cost, to achieve victory in a given war. The veteran and his or her spouse must seriously work hard at keeping the relationship, finding or even inventing ways to have fun together and consciously rebuilding the relationship.

Chapter IV
WHAT IS THE COMBAT VETERAN LIKE?

We all have idiosyncrasies that developed as we each grew up in our family of origin. We take on the family's tendency to be neat or sloppy of person or housekeeping, to eat with finesse or with slovenly habits, to have an orderly home where everything is in its place or a messy house where nothing is put away, drawers left open, clothing on the floor, etc. Accordingly, anyone serving in the military develops characteristics that are different from those of non-service people, and intense enough to become fixed. There are characteristics the veteran comes home with which can be either directly or indirectly connected to combat traumas, and which have made him or her a different person. This chapter discusses many of those characteristics

Numbing Out

The numbing out in the veteran, discussed in the previous chapter, seems to the spouse and children of a veteran as though the veteran doesn't really care about them. The veteran doesn't respond to them with any visible emotion, except perhaps storming out to be alone somewhere. Most of us don't realize that we judge another person's feelings for us by what kind of response we can get by pushing the person's emotional buttons. The veteran may be passionate, but in emotionally charged situations that appear to be a personal attack, he or she can only observe and take what appears to be impulsive action without emotions. It is like his or her "buttons" don't work anymore. The person trying to push the veteran's buttons wrongly deduces that the veteran just doesn't care about them.

Avoiding confrontation that would bring on anger and keep the veteran from successfully dealing with debate about relationship, job, or other necessary parts of his or her life, becomes an art. The veteran subconsciously invents ways to avoid anger and any feelings, and particularly events that would initiate a flashback. Before I became conscious of my avoidance behaviors, I managed to avoid going to funerals with one trumped up excuse after another. I managed to miss my Grandmother's funeral (by insisting on a traditional honeymoon that same week), my step-children's Grandfather's funeral (by flying into the wrong airport), and even my own Mother's (by deciding she wouldn't have wanted me to be devastated).

That emotional shut-down which the veteran learned when in the attack mode and when anger is primary, also manifests itself in some veterans as abrupt avoidance of people who are angry or aggressive toward the veteran. Better to shut down and avoid than to respond with the veteran's own aggression, particularly when the other person is spouse, children, boss, or fellow worker, and especially when the veteran is afraid he or she can't control anger in aggression.

Anger

A constant part of the numbing out in combat is the anger that drives the veteran on, to get through the horrors of combat. As discussed in previous chapters, that anger is an automatic response to any challenge that provokes need for performance done in the face of threat that would normally evoke emotions, but where emotions would just get in the way. After returning home the veteran meets every challenge, whether relationship, job or even sports, with anger when provoked with a pressure that devolves into his or her normal combat mode: full attention, full intensity, full anger, immediate action, and no emotions.

Most veterans are afraid of their anger, having done things like killing another human being in the middle of combat's anger drive, things certainly not acceptable in

60

civilized society. So most veterans know they have had no emotions about killing ever since that lack of feelings was trained into him or her as a recruit. Veterans are afraid of what they might do in anger, when for an unconscious instant he or she might revert to combat behavior. The veteran rarely is provoked to those actions, but the veteran is still afraid he or she **might** be. In therapy, the veteran has to talk about and re-live all those experiences he or she has been avoiding thinking about for years, and so getting into therapy may be a harrowing pain. But it is more than worth it even if the only gain you would get would be the realization that **you are no longer capable** of doing the horrendous things you once did, without emotion.

Throughout this chapter IV, I ask the reader to understand that the phrase "because of PTSD" may mean the veteran didn't have a choice whether or not to have those habits and characteristics necessary to survive in combat, but that doesn't mean that the veteran can't change. You have choices. You can stay as you are, suffering the endless recurring nightmares and flashbacks, and being out of step with everyone else's reality, or you can chose to heal, step by step, one habit at a time. The veteran with PTSD, *because of the PTSD,* is prone to have unsolicited anger engulf him or her simply because he or she is trying to respond to pressure from an everyday relationship glitch or from a job or sports challenge, but that can be fixed. With therapy, that anger can be channeled into more socially acceptable behavior.

Addiction

Sometimes, a veteran learns how to become a workaholic, another form of numbing out. The veteran works around the clock, not allowing time for emotional threats, thereby avoiding those emotional situations that make the veteran so unhappy. Who can criticize him or her for working extra hard and providing for the family?

A workaholic is often called an addict. Being workaholic **is** an addiction, that is true. The numbing out which the combat veteran learned in order to survive is no longer needed for most experiences and activities after discharge, but the veteran has a deeply rooted habit to avoid feelings in any intense situation, and so he or she finds it easy to slip into addictions. As anyone in a 12-Step Program knows, the basic function of addiction is to numb out feelings and to avoid thoughts about a person's behavior. And anyone in a 12-Step program also knows that help is there for anyone to get back control of his or her life and find change. Scientists say that addiction is learned behavior[18], and that learned behavior can be changed.

Another source of addiction is the heavy drinking most veterans did and perhaps still do. It is again a sort of double whammy the veteran gets, not only burdened with PTSD which takes much work to deal with, but unknowingly slipping into addictions which are now additional burdens that have to be dealt with, often with tragic affect to the veteran and to his or her loved ones, jobs, friends, and health. Many combat veterans with addictions find real healing by entering both 12-Step programs and also at the same time starting VA Mental Health therapy.

I have met many veterans of different wars, and the eighteenth century myths about "crazy" people are well and hardy –many veterans, when prompted even with explanations and genuine, caring TLC, can't accept needing mental health clinic help. It is too bad that the VA Mental Health Clinic can't change the name to something like "PTSD Reversal Clinic." As we saw in the previous chapter, PTSD **is** a physical **as well as** psychological affliction.

Impulsiveness

Another characteristic of most combat veterans is impulsiveness. In combat, for every deadly situation thrust upon him or her, the veteran had to make a split second

---

[18]   New York Times newspaper, Vol.CLVIII...No.54,637 (April 6, 2009) "Brain Researchers Open Door to Editing Unwanted Memory"

analysis and then immediately act on the quick decision, seemingly the first impulse entering his or her head. As a recruit, his training tried to set up reflexes for combat situations to cut the reaction time even shorter. The veterans who survived combat learn to trust their impulses, making quick decisions and taking action immediately, usually without any consciousness of that process. Conscious thought takes time; conditioned reactions are quicker. To the non-veteran, that self-confidence comes across as aggressive arrogance

When the veteran comes home, it shouldn't be surprising that the veteran carries that impulsiveness throughout his or her life. That's not always good in civilian life with its complicated, unpracticed situations and often with poor information that needs analysis, confirmation, negotiation, and evaluative and collective thought with spouse or children before making decisions. The incoming artillery fire has ceased; the bombs aren't falling anymore; the rifle and machine gun fire is silent. A buddy isn't lying there, wounded, bleeding to death. There isn't really any life-threatening situation demanding urgency to face in daily living. But the veteran is caught in the web of his combat conditioning and must keep trying to solve the problem quickly, stay calm, keep going........

A loved one may not be able to comprehend how bewildering it is to the veteran to keep trying to change from not just one of these combat driven conditionings, but from many of them which we are speaking of throughout this chapter, all interactive and webbed together, continually dragging him or her back into combat-driven behavior. But it happens, over and over.

Habit Personality

There are characteristics not directly connected to combat traumas, habits that the veteran may have developed which are different than civilians expect. One that most veterans come home with and which could be offensive to others, is the veteran's demand for verification of information. As noted in the letter to the *Ozark News* mentioned at the end of Chapter I, intelligence of combat situations was a daily concern in combat. The combat veteran gets used to needing to know all details of where information came from. Intelligence Unit soldiers made a career of learning how to ascertain veracity of information. The veteran coming home seems almost belligerent in wanting to know how you learned about a piece of information, not because the veteran really needs to know, but it is his or her make-up from combat to try to know the validity of any information.

It is sad that in our instant-information culture, we've lost a sense of what is verifiable and what isn't. We talk about the news as though we were personally there and know it for a fact, not indicating any clue where the information came from. We talk about the weather as though we are meteorologists and know these facts we express. We talk about politicians' characteristics as though we are analysts who have studied these people and know them intimately. We are ready to argue vehemently about these things that we "absolutely know for a fact." But the veteran can't just suddenly drop back into that culture. He or she irrevocably thinks in Eskimo[19], and therefore needs to know the source of information and to say where he got his. It's a part of his thinking, after being in combat.

---

[19] The veteran could be comfortable conversing with an Eskimo if he or she understood the languages. The Eskimo languages don't have temporal conjugations (past, present, future) the way most languages do, because time isn't as important to an Eskimo as is the veracity of information. The Eskimo language **does** have three conjugations that they use in speaking, but they are: 1) I personally experienced it, saw it, felt it, etc.; 2) I heard about it from someone else, or deduced it from available information; and 3) I'm being poetic and using my imagination to create a story or image of something that may not exist. When forced by hunger to go out on the ice in a freezing blizzard after a dangerous polar bear, the veracity of the information had better be certain!

The veteran in combat has had life experiences so intense and continuing and so different than those his family knew in his childhood that he or she is markedly changed with habits too strong to easily change again: eating too fast (the artillery shelling might start anytime); buying more food than can possibly be eaten before spoiling (don't know where or when the next meal will be); being so focused on a project that meals and even sleep are ignored (full speed ahead, make decisions, keep going, no matter what's happening around you); being able to fall asleep easily, but never sleeping soundly (may have to move out any time); not fastidious of personal hygiene (the only water we have is for drinking, and there's not enough of that), and so forth.

The combat veteran's natural outgoing or solitary personality is greatly exaggerated by his or her experiences in combat. Those who always were loners don't like crowds, don't even like being in noisy social situations, and are difficult to get into a conversation. On the other hand, those who were outgoing like to talk and often like dialog and debate. The gregarious veterans don't like to be alone and will take abuse just to be with others. They often are sports fans in the extreme, going regularly to games with buddies, expressing camaraderie, all acting as one, cheering the team on.

They seek to relive the intense camaraderie they had in combat, where gross humor and any sort of kidding or practical jokes was all the levity a soldier had, and one more option to hide real feelings. When veterans get together in, say, a reunion or a WWII group therapy, they are back in the banter and kidding mode, part of why they feel comfortable in such settings.

The loner veteran is the more typical in civilian life, at least to appearances. The veteran may yearn to be back with his or her buddies, being a part of an "extended" family of military comrades, to be gregarious and have connection with people, but the veteran finds it difficult to fit into the "you scratch my back and I'll scratch yours" mind set of the civilian world.

Army training was all about self-sufficiency. No one helped you make your bed or shine your shoes or do your laundry or take your turn for KP duty in boot camp. You learned to do everything alone, but in the company of many who were also taking care of their own needs. In combat, the same applied, except that when you saw someone in trouble, you automatically and immediately stepped in to help, to silence an enemy machine gun or to help lug an impossibly heavy batch of Bangalore torpedoes, etc. You never asked for help, and no one asked you for help; it was just an unwritten code to go alone unless someone happened to step in to help, and if you saw someone who needed help, you just stepped in to help. When in a threatening combat situation, the combat soldier didn't have time to look around for help, but had to act as though no one else was there to help him or her, and to take immediate action, as always.

So the veteran, now home, finds it difficult to ask for help for any need. He or she will stoically go it alone in any endeavor, and is overjoyed if someone asks for his or her help. To others, the veteran's do-it-alone style may appear to be unfriendliness, and when in a club situation where for instance the banquet preparation needs to be done, the veteran, often a leader (by example) in a military situation, often will desperately want to help and could lead a team of helpers, but he or she doesn't know how to tell people what to do in that civilian situation. In combat, you never have to tell someone how to do his or her job.

Cleanliness and Hygiene Disappeared

It follows that habits of cleanliness and hygiene disappeared along with other civilian characteristics. The veteran was often oblivious of his or her living in what could be a very messy environment, because the combat soldier lived for months or years in desolation and disorder and rarely stayed in one place more than a few days. The concept of homey orderliness (put the toilet seat down, honey) just didn't exist for him or her in the combat environment. Even clean clothes and shined shoes were anathema for the combat veteran, because he or she seldom had time or need for that, and how do you polish muddy shoes when you spend days living and sleeping in mud –it's better to get in the habit not to think about it.

After living for so long without routine in their lives, combat veterans can bewilder the non-veteran by being able to "suffer hardship" for no apparent reason, such as just deciding to sleep on a towel on the floor at work after working too late and being terribly tired. In fact, there was no hardship to the veteran; it just seemed logical to go to sleep when that tired, and sleeping anywhere in any environment was no big deal.

Even personal hygiene was a quite different world for the combat soldier. During WWII combat, many units were on line without relief for months sometimes, like the 218 days we in Company B were on line, so you just got out of the habit of anything approaching cleanliness. After VJ-Day, some green replacements for our then latest officer casualties (officers didn't last long in WWII combat) tried to get us to brush our teeth. We'd done little washing of ourselves or our clothing during combat, or shaving, since we often didn't have enough water even to drink. We eventually just forgot about those things, and, of course, we didn't carry razors or toothbrushes in our combat gear. The new 2nd Lt. was not very popular, making us fall out for tooth brushing every morning after he took the trouble to get toothbrushes sent in.

The rear echelon did try several times to promote good hygiene. Once, the order came up to the town we were fighting in that water was being sent up to the front and we were ordered to shave (because, believe it or not, the order said we were looking pretty scroungy—it may have been because the battalion or regimental commander, who we seldom saw in combat, had visited the front lines and didn't like what they saw). However, that was one of those times when we didn't have enough water to drink, so our company CO countermanded the order, letting us get drunk on all the fresh water we could drink, a real treat.

Likewise, in the early days of the winter push into Germany, we once were rotated to the rear in Holland for a hot shower and clean clothes. But since we had been living outside in the cold for so long, the attempt at hygiene caused widespread colds, and a few men with pneumonia had to be hospitalized. The rear echelon didn't try again to experiment with giving us cleanliness! That illustrates again how far apart the reality is for those in combat and those who aren't. Those who have not been in combat simply cannot know what it is like, and cannot think in terms of the realities of combat.

The living conditions were in themselves traumatic, and the physical stress to carry the equipment, weapons, ammunition, food and overcoats with us everywhere everyday, was a lot to ask of us. We of course had no place to stash anything "while we were away in combat." We never were away **from** combat: what we needed or wanted just had to be carried with us everywhere. And the only way to move at the front was our own tired feet; there were no trucks to carry us in WWII close action combat. The few times we were trucked between engagement areas, it was a treat to be momentarily out of combat. Combat was not just physically overwhelming, but the endlessness of it was traumatic, both to our bodies and to our psyches.

Hyper-Alertness

A combat veteran is usually hyper-alert, especially when he or she first comes home. Those who survived intense combat have an intense alertness, trying to see and evaluate everything going on around them. When they first come home, they never feel comfortable in a crowded room, unless sitting in the back of the classroom or having their backs to a wall so that they can't be surprised by anything unseen behind them.

The combat veteran usually awakes fully alert and already moving. Sometimes the body is in action before the mind is fully awake, which can show some bizarre and even comical behavior. When I was a student at the University of Michigan, some non-veterans in the dorm at Ypsilanti learned how to pick my door lock, would sneak into my room in the middle of the night, and then on cue roughly shake me and shine a blinding flashlight in my face. Although not awake enough to evaluate what was happening or even to remember it had happened before, I'd instantly charge them, swinging and tackling. They thought it was great fun, and somehow managed to not get hurt.

But the habit of being hyper-alert also has some advantages. For instance, it makes the veteran a better driver than the average person. The veteran is observing the driving environment and easily seeing everything in his vision, especially anything moving, with an intensity that the average person takes many years to learn; many don't learn that at all. The veteran's hyper-alert intensity makes a single-purposed focus that eliminates conversation. (Why can't we teach that to our children!)

Some veterans really enjoy driving, particularly long distance, high-speed trips, finding that it gives him or her a little of the combat feeling of being hyper-alive in a risky environment, and especially when alone, so no one is trying to compromise his or her focus. The combat veteran doesn't like to drive behind another vehicle, especially if he or she can't see traffic up front, and will maneuver aggressively to pass other vehicles, simply to not be behind them..

Survival Personality

In general, the personality characteristics that a veteran had when he or she went into combat tended to either help the veteran survive or to not help, just as children tend to develop personality characteristics which help them to survive family and school dysfunctions. Those

69

characteristics learned by the veteran to survive combat become reinforced and more exaggerated as the veteran experiences repeated combat action and the veteran feels more confident in expressing those characteristics. If caution helped him or her to survive, then the veteran becomes stubbornly cautious (call it being General Montgomery); if aggression helped survival, then the veteran becomes viciously aggressive (call it being General Patton).

When the veteran comes home, his actions then become all out of proportion to situations, which is extremely frustrating and puzzling to the civilian trying to understand him or her. The veteran's behavior just doesn't seem to make sense. Especially disconcerting to civilians is the intensity of the veteran, and the seriousness at times and the meaningless banter at other times.

Lawlessness

The biggest social problem with returning combat veterans is lawlessness. It is one of the most difficult to fix, particularly if not caught early after discharge and felonies are committed. Incarceration only exacerbates the problem and draws the veteran more deeply into his affliction.

I was asked once where we got all the booze we imbibed in combat and after. The general sense of the question was that we shouldn't have been *allowed* to become drunkards. But understand that there were no laws or regulations governing us in combat other than to follow orders, and except for the law of survival –that is, anything goes if it keeps us alive. It was anarchy in its most primitive form. During combat any booze we found (and some guys dedicatedly looked for it between combat actions), we took what we wanted. We scrounged for food and took what we wanted. We sometimes looked for women and took what we wanted.

One GI carried a German sewing machine on his back for weeks until he could send it home to his poor family. Everyone carried whatever he wanted. If he wanted to carry a coin collection he found in a dentist's bombed out office, instead of food or ammo or cigarettes, that was his business. If his eccentricity got him killed, that, too, was his business.

I remember one time several of us on a patrol came across a bombed out sheet-glass factory, with hundreds of sheets of glass, all sizes, standing on edge throughout the factory, many still unbroken. Can you imagine the boyhood dream of being able to smash hundreds of panes of glass without consequences? We wasted a lot of ammo there, but it felt good.

Lawlessness was not something we sort of impishly grinned about while, say, shooting up a glass factory. We simply got in the habit of making our own choices; in our judgment, if the contemplated action wouldn't get us killed, why not do it. For example, once in occupation in Bavaria, when the demolition guys planned to blow up a German ammo dump outside the town we were occupying, notices were carefully given to all the townspeople to stay indoors in basements on the scheduled morning, and to not go outside until after the many sections of the ammo dump were blown up, one at a time. We GIs also were ordered to go into homes that had basements and to stay there until all the ammo sections were detonated.

Two of us were curious what it would look like to see big caches of ammo go up, so we stayed on the basement stairs near the top, looking out the door to watch. Unfortunately, the demo guys screwed up and the whole dump went up at once, creating an eruption that showered the entire town with debris, like a volcano eruption. The two of us were so startled that we jumped down the whole flight of stairs, tripping each other and rolling to the bottom like in a bad movie.

If we got hurt from not following orders, that was just our own business. The only time officers butted in was if the actions of one guy put others in danger –that was not allowed. Usually the guys who were put in harms way because of another's bad judgment made the call. For example, once when my squad was to go on a nighttime combat patrol early in our introduction to combat, I tried to go but was ordered not to because I wore glasses that could reflect what little light was around and give the patrol's location away. Although mortar squads were usually busy at night doing their own thing and seldom sent out on patrols, our familiarity with carbines and 45 caliber sub-machine guns and pistols, our normal side arms, made us useful for night patrols. I was depressed to be singled out not to go with my squad and argued to no avail.

The lawlessness was firmly entrenched in our psyche after living like that for a year or two. The corollary concept is that we forgot to consider consequences of our actions. That is, to analyze a situation and realize that if we did what we were ordered to do, the consequences could be getting killed, so we just stopped looking at consequences. It was understood that we might be killed, so why even think about it. The habit, then, that the combat veteran brings home is to simply block thinking about conse-quences of actions, not very acceptable to society, where one of the first things we try to teach our toddlers is consequences of actions.

Corollary to that was the habit a veteran brings home of not thinking about the future, of not planning ahead; in combat, you not only don't know what's going on or even where you are, you really don't **want** to know; you are too busy with the details of fighting that you don't really care about anything except surviving the next few minutes. Thinking about the future, with its probability of not existing, was a booby-trap which the combat veteran avoided completely. That habit, brought home, is avoidance of thinking not just about future plans, but even everyday plans of what to do tomorrow or what house repairs or fixings need to be done or who's birthday is coming up. The combat veteran seems in a stupor about what's going on around him or her, drifting from one day to the next, just acting compulsively to any suggestion he or she seems to like.

Another aspect of the veteran's lawlessness is the habit of questioning rules (as different than combat orders) –rules weren't to just be blindly followed –they were to be analyzed and if safe, and if there was something in it for the veteran, then to follow them was O.K.; otherwise, there wasn't any point in following the rule. Often there wasn't time to evaluate a rule –simpler to just ignore it. We of course knew right from wrong, but the habit of not **seeing** anything wrong with making our own decisions, even to lawless behavior, resulted in hometown surprise, surprise – police sirens--you got caught and have no alibi, only deep confusion.

Trying to fit into a society that wouldn't allow lawlessness and not thinking about consequences was perhaps the most difficult adjustment for a veteran to deal with in coming home. After VJ-Day, nothing changed, but I won't go into much of what we did in occupation because that is the hardest for me to try to forget. You say that war justified what we did during combat, but what about after that excuse ended? Was the excuse any different during or after, or were we who survived just bad people.

Poor Self-Esteem

A combat veteran may come home a lot more of a loner than when he or she left. In combat, a soldier learns that what he used to take for granted from childhood as the real world just isn't true. Contrary to childhood training, the world isn't a safe place; you can be snuffed out in an instant. All people aren't good; you no longer can assume that the next person to come along could be a friend. You learn that the only person you can really trust is yourself.

Also, if you start to feel that your buddies are the only real friends you can count on, you learn the hard way that some will die tomorrow or the next day, or the day after that. You may just stop making close friends to avoid the pain of losing them. And you learn that if they save your life (as they often do), you can never repay them. You feel a haunting debt that never goes away.

You no longer can see yourself as a good person, not after doing the things you did. Sure, the war and its killing was something that had to be done, as the veterans are often told, but the combat soldier feels that by actually killing men and even women and children, and doing those other horrific things, especially to women, how can he ever think of himself as a good person?

The world of people he tries to interrelate with, the world of safety and friends, he can't communicate with anymore. As noted in the previous chapter, his world of combat realities is an entirely different world than the world that people at home live in, and the civilians just can't communicate easily with the veteran, either. In letters while in combat, both the veteran and the loved ones may start to see the disconnect of those worlds, and when the veteran comes home, that disconnect still confuses him or her. Communication seems impossible and fraught with mines and booby traps.

If the two people in a relationship are ignorant of the factors discussed is the previous chapter (Two Worlds), a blame game might get started, where each of a couple start blaming the other for the disconnect. This is where TLC is most effective and also the most difficult, without VA counseling help.

The combat induced personality is exacerbated by other minor emotional traumas that happen: the soldier doesn't get mail from someone who should care; news from home is traumatic and the veteran can't do anything about it. The only strands of human connection we had with reality, with the world of home, were the letters we wrote and received. Christmas of '45, when we were not home yet, was so painful to most of us that we had to get drunk on Christmas Eve to try to blank out the hurt.

The worst, though, was getting "Dear John" letters when the sweetheart or wife was the only dream we could have. I had irrevocably scratched my sweetheart's name, *Peggy*, on the 60mm mortar barrel I carried everywhere in combat. I got a Dear Frank letter from her, telling me she had found someone else. I received it the day before the attack on Welz, and that mortar barrel was left somewhere in the beet fields of Germany's Cologne Plain, when I was wounded.

**

But the craziest part of all this is that although the veteran survived the war by suppressing emotions and learning to act impulsively by suppressing any conscious thought, when he or she carries those habits into his new life and relationships, there is a horrendous price to pay. ***Those habits will kill him or her at a younger age than non-veteran peers.*** WHY? Because when a person suppresses emotions or even thoughts about important decisions continually over a period of time, that habit also compromises the person's immune system. Scientists have proven that suppressing emotions also suppresses the immune system. By suppressing his or her emotions, that is, numbing out, the veteran will be plagued with endless medical problems both small and serious throughout life.

On the other hand, if he or she instead lets out the pent-up anger and emotions, society doesn't accept the resulting behavior and sooner or later, he or she is incarcerated for unsociable actions. The combat veteran can't win, without serious help and counseling and/or prescribed drugs.

**

Finally, let me again reference James Jones' book ***WWII,*** wherein he's talking about the factors that seem to cause many veterans to get in the fast lane for excitement. They drive too fast, and take right of way when it isn't theirs. They take short cuts with time, trying to fit too much into each day. They J-walk without looking, and don't bother to read road signs. They have no patience with rules and regulations, let debts go unanswered or don't pay taxes on time. They do risky sports like motorcycle riding or sky-diving and take jobs like oil well digging. They like to carry guns or have them, virtual collections, in their home. They try to find intensity in people, companions, jobs, all of life. They try to go "full speed ahead, damn the torpedoes." It's all to recreate a vague war memory of when it felt good to be alive.

James Jones:

...even those of us who hated (combat) found it exciting, --sometimes. That is what the civilian people never understand about their returning soldiers, in any war…. They cannot understand how we could hate it, and still like it; and they do not realize they have a lot of dead men among them, dead men who are walking around and breathing. Some men find it hard to come back from their evolution of a (combat) soldier. Some never come back at all, not completely (Jones 45).

ADDENDUM

Finding Communication

To all the spouses of combat veterans, those who stayed with them and didn't quit, I salute you –I've met many WWII combat veterans who have 50 and 60 year marriages, and over those years, the veterans became re-acultured and socially acceptable, thanks in large part to their spouses' devotion. It's an interesting thought that those WWII veterans who had devoted spouses have survived to ages 80's and 90's, while many of those who didn't, died younger.

I've learned as I've grown older that the infirmities of growing old are not as depressing as physical handicaps were when I was younger. Just knowing the joy of being alive and rejoicing in all that life has to give me, makes me a happier veteran than I was as a new veteran and a young man. Many friends, loves, and family have helped me to that place.

My advise to you all, in summation, is to try to accept any veteran coming home, not on your terms, but on his or her terms, even when it seems to make no sense or is wholly repugnant to you. <u>Every veteran has a reason for what he or she does or says, which makes sense to some real part of him or her, but not necessarily to you.</u>

I sincerely hope that this book will help non-veterans to have more flexibility to tolerate undesirable actions of veterans –after all it was your tax dollars that sent them into war. Remember, the veteran is still, for many years, living in the world of combat meaning, not your world of order and civility. Your biggest challenge is to keep trying, through loving communication, to bridge the gap between your world and the combat-generated one that is his or hers, maybe only for a while, maybe for always.

And nobody should ever believe it will be easy

Lest we forget why we need to try to understand our veterans, let me quote a poem my sister sent me once, a few years ago:

IT IS THE SOLDIER

It is the soldier, not the preacher, who has given us freedom of religion.

It is the soldier, not the reporter, who has given us freedom of the press.

It is the soldier, not the poet, who has given us freedom of speech.

It is the soldier, not the campus organizer, who has given us freedom of assembly.

It is the soldier, not the lawyer, who has given us the right to a fair trial.

It is the soldier, not the politician, who has given us the right to vote.

It is the soldier, not the statesman, who has given us a government of the people, by the people, and for the people.

--Our America!

(author unknown; ending changed by Francis Resta)

A Call For Action

One thing I also learned in 1990 from the Howe Avenue California Veterans Center counselors was that a significant percentage of the inmates in California prisons at that time were combat veterans from WWII, Korea, and Vietnam, but were receiving no VA or Cal State PTSD treatment benefits. I asked why weren't they being given the same treatment that I was getting. I was told that the veteran's agencies, both State and Federal, had tried but the California prisons would not allow it because they were afraid that the prisoners who were combat veterans would all then want retrials based on their crimes not being their fault, but because of a treatable disease. So, tragically, there is no therapy for incarcerated combat veterans, the ones who need it the most.

I wish to make a plea: If you know anyone who is a politician or who has a close connection to one, tell him or her the story above. Let's hope that somehow the returning combat veterans from Iraq and Afghanistan won't be needlessly incarcerated with no hope for help, simply because they are combat veterans who are socially unacceptable through no fault of their own. It would seem fitting that a great and respected attorney, somewhere, would be only too anxious to research the facts, mount a class action suit for those incarcerated veterans, and rescue them.

Appendix A - VETERANS BENEFITS
Data applies to year 2008

As many of you know or could guess, there are many levels of government which provide veterans benefits; You may **not** know that benefits are not only for veterans, but also possibly
> ✓ for a Veteran's dependent, or a surviving spouse, or child
> ✓ or <u>parent</u>, even,
> ✓ or active duty military service member,
> ✓ or members of a Reserve or National guard.

Benefits are listed here by the various levels of government, with California shown as an example of State. You can either find them in the telephone book under government listings, or also on Internet by just goggling on <u>Veterans Benefits</u> at whatever government level you wish, e.g., U.S. Veterans Benefits or California State Veterans Benefits.

At the Federal Level, the U.S. Department of Veterans Affairs (VA) will provide a veteran with:
- A monthly <u>compensation</u> payment if disabled, and including benefits to dependents and survivors
- A monthly <u>pension payment,</u> if disabled, over 65 and with little income
- Many <u>health care services</u> at a VA hospital or clinic at reduced cost or free for service-incurred problems
- <u>Pharmacy benefits</u> at reduced cost or free for some veterans
- Free <u>PTSD health care</u> for combat veterans. (PTSD is Post Traumatic Stress Disorder)
- <u>Vocational Rehabilitation & Employment</u>
- <u>Transition Assistance</u> to help veterans and families cope with readjusting back into our communities for both disabled and just plain needy veterans

81

- Education and Training under various bills including the various versions of the G.I Bill and the latest, new, huge 2008 version
- Home Loans at reduced rates and with U.S. Government guarantees
- Life Insurance at reduced costs
- Veterans Preference for Federal Civil Service Jobs
- Burial benefits and services

At the State Level, the California (for example) Department of Veterans Affairs provides many benefits and services which add up to a lot of help for veterans and their families:
- California tuition waivers for Veteran's dependents
- California Motor Vehicle registration fee waiver
- Disabled Veterans Business Enterprise Opportunity
- Disabled Veterans License Plates and some other free license plates
- Veterans Home of California access at Yountville, Barstow, and Chula Vista
- Veterans Preference in California State Civil Service exams
- State Parks and Recreation Fees Waivers
- Veterans Claims representation
- Employment and Unemployment Insurance Assistance
- California Veterans Farm and Home Loans
- Business License, Tax, & fee waiver
- Non-Resident college fee waiver
- Fishing and Hunting licenses
- Property tax exemptions
- State of California Veterans Cemetery

Appendix B – HOW ONE GOES ABOUT
GETTING VETERANS BENEFITS
Data applies to year 2008

Appendix A is an impressive list of benefits for veterans by the state and federal governments. But how does a veteran go about getting them?

At the county level, the **County Veterans Services Office** in each county is the most important link between veterans and their benefits. It provides much assistance to veterans and their families by helping them find their way through the labyrinth of offices and provisions to receive the federal and state benefits, quickly and positively. I can personally attest that the Yolo County Services Office in Woodland, California, was persistent in communicating with the other agencies and unbending until my claims were answered.

At the city level, most cities have agencies such as the Senior Center in Davis, California, which has a Veterans Affairs Office on Mondays, again to help application for the many benefits available to veterans and their families.

Most cities also have groups or agencies such as the Senior Citizens of Davis, California, which will help veterans and their families find assistance for almost any problem.

And then, in addition to levels of government, there are veterans organizations which not only provide support groups for veterans, but also will assist veterans and their families get the many benefits available. Veterans are more inclined to seek help if it's offered as part of the bonding of their own kind which all veterans feel and trust.

Here is a partial list of veterans organizations in my area:
- The American Legion
- The "Help Paralyzed Veterans" Organization
- The Blind Veterans Association
- The Marine Corps League
- The AMVETS, American Veterans, which pick up tax deductible donations, in addition to providing many other kinds of help for veterans
- The Pearl Harbor Survivors
- The Disabled American Veterans (DAV) and the DAV Auxiliary. Quote from a DAV flyer (probably circa 2004):

> "According to the Dept. of Veterans Affairs (VA), there are 1.2 million female veterans, making up about 4 % of the veteran population. And those numbers continue to rise. The federal Government predicts that by the year 2010, nearly 10% of the veterans will be women.
>
> "Women veterans use their earned benefits at lower rates than do their male counterparts. Therefore, we must ensure there is proper outreach to women veterans and that they are not hampered in getting the benefits and health care they need."

- **And** --The Veterans of Foreign Wars with chapters in many cities and suburban areas of large cities, publishes a monthly VFW magazine with articles on current issues important to veterans.

Each Veterans organization provides its own versions of help and assistance both to veterans and also to the general community in many ways such as sponsoring a Boy Scouts troop, in addition to providing a support place for veterans to meet with their own kind.

The County Veterans Assistance office can help you find veterans organizations that you might be interested in, both local clubs, and also local   branches of the national organizations such as those listed above.

And, finally, almost **every active and deactivated military unit has an association** that a veteran can join. The associations' reunions create support groups for veterans, which provides important social bonding for the veteran and ways to keep track of his or her old buddies. Most unit associations will try to help any of their member veterans and families with a benefits problem. You can get to such associations by contacting the U.S. Department of Defense.

TABLE OF FIGURES

REFERENCES

The film: *The Big Red One*...............................3, 48

Ken Burns' documentary film: *The War*
...............5, 6, 8, 12-13, 22, 34-35, 41, 54, 56

James Jones: *WWII, A Chronicle of Soldiering*
...............................ix, 10-11, 35, 41, 76

Letter by the author to the Editor of the *Ozark Notes*..16-18

Other references to *Ozark Notes*...............15, 18, 28-30, 64

Alva Noë: *Out of Our Heads*...........................52, 55

Poem, author unknown: "It Is The Soldier"...................78